STRUCTURE AND BONDING is issued at irregular intervals, according to the material received. With the acceptance for publication of a manuscript, copyright of all countries is vested exclusively in the publisher. Only papers not previously published elsewhere should be submitted. Likewise, the author guarantees against subsequent publication elsewhere. The text should be as clear and concise as possible, the manuscript written on one side of the paper only. Illustrations should be limited to those actually necessary.

Manuscripts will be accepted by the editors:

Professor Dr. *J. D. Dunitz* Laboratorium für Organische Chemie der Eidgenössischen Hochschule
CH-8006 Zürich, Universitätsstraße 6/8

Professor Dr. *P. Hemmerich* Universität Konstanz, Fachbereich Biologie
D-7750 Konstanz, Postfach 733

Professor *J. A. Ibers* Department of Chemistry, Northwestern University
Evanston, Illinois 60201/USA

Professor Dr. *C. K. Jørgensen* 51, Route de Frontenex,
CH-1207 Genève

Professor *J. B. Neilands* University of California, Biochemistry Department
Berkeley, California 94720/USA

Sir *Ronald S. Nyholm*, FRS †

Professor Dr. *D. Reinen* Institut für Anorganische Chemie der Universität Marburg
D-3550 Marburg, Gutenbergstraße 18

Professor *R. J. P. Williams* Wadham College, Inorganic Chemistry Laboratory
Oxford OX1 3QR/Great Britain

SPRINGER-VERLAG

D-6900 Heidelberg 1
P. O. Box 1780
Telephone (06221) 4 91 01
Telex 04-61 723

D-1000 Berlin 33
Heidelberger Platz 3
Telephone (0311) 82 20 01
Telex 01-83 319

SPRINGER-VERLAG
NEW YORK INC.

175, Fifth Avenue
New York, N. Y. 10010
Telephone 673-2660

STRUCTURE
AND BONDING

Volume 11

Editors: J. D. Dunitz, Zürich
P. Hemmerich, Konstanz · J. A. Ibers, Evanston
C. K. Jørgensen, Genève · J. B. Neilands, Berkeley
Sir Ronald S. Nyholm†, London · D. Reinen, Marburg · R. J. P. Williams, Oxford

With 58 Figures

Springer-Verlag
Berlin Heidelberg GmbH 1972

ISBN 978-3-540-05830-4   ISBN 978-3-540-37473-2 (eBook)
DOI 10.1007/978-3-540-37473-2

The use of general descriptive names, trade marks, etc. in this publication, even if the former are not especially identified, is not to be taken as a sign that such names, as understood by the Trade Marks and Merchandise Marks Act, may accordingly be used freely by anyone.

This work is subject to copyright. All rights are reserved, whether the whole or part of the material is concerned, specifically those of translation, reprinting, re-use of illustrations, broadcasting, reproduction by photocopying machine or similar means, and storage in data banks. Under § 54 of the German Copyright Law where copies are made for other than private use, a fee is payable to the publisher, the amount of the fee to be determined by agreement with the publisher.
© by Springer-Verlag Berlin Heidelberg 1972
Originally published by Springer-Verlag Berlin · Heidelberg · New York in 1972

Library of Congress Catalog Card Number 67-11280.

Contents

Contents

The Chemistry of Complexes Related to *cis*-Pt(NH₃)₂Cl₂. An Anti-Tumour Drug

The Chemistry of Complexes Related to cis-Pt(NH$_3$)$_2$Cl$_2$. An Anti-Tumour Drug

A. J. Thomson
School of Chemical Sciences, University of East Anglia, Norwich, Norfolk, England

R. J. P. Williams
Wadham College and Department of Inorganic Chemistry, University of Oxford, Oxford, England.

and in part

S. Reslova
Department of Biophysics, Faculty of Science, Charles University, Prague, Czechoslovakia

Table of Contents

A. J. Thomson, R. J. P. Williams, and S. Reslova

I. Introduction

The starting point of this article is the discovery by *Rosenberg* and his collaborators of the activity of certain platinum compounds as anti-tumour agents (*1*). Here we wish to introduce the chemistry of these agents, which will be relatively familiar to inorganic chemists, but not to chemists and biochemists of other disciplines. At the same time we wish the inorganic chemist to see the relevance of the chemistry to biological fields.

The important observations of Rosenberg were reported in 1969 (*1*). Then it was shown that

$$cis\text{-}Pt(NH_3)_2Cl_4,$$

$$cis\text{-}Pt(NH_3)_2Cl_2,$$

$$PtenCl_2,$$

$$\text{and } Pt(en)Cl_4,$$

where en is ethylenediamine, inhibited sarcoma 180 and leukaemia L 1210 in mice. Earlier *Rosenberg, Van Camp,* and *Krigas* (*2*) had shown that a very similar group of compounds were effective in the inhibition of cell division in *Escherichia coli* (Table 1). Subsequently *Reslova* (*3*)

Table 1. *Effects of certain group VIII transition metal compounds in producing elongation in Escherichia Coli after 6 hr. of incubation in synthetic medium (4)*

Metal compound	Effective conc (μg/ml)	% Filaments	Elongation [a] (estimated)
$(NH_4)_2PtCl_6$[b])	1—20	100%	25—100 X
$RhCl_3$	30—100	75%	5—25 X
$(NH_4)_2RhCl_6$	20—30	75%	5—20 X
K_2RuCl_6	40—50	50%	5—20 X
$K_2RuNOCl_5$	40—100	50%	5—10 X
RuI_2	20—30	25%	10—20 X
$RuNOCl_3$	20—40	25%	4—6 X
$RuBr_2$	10—20	10%	10—20 X
$RuNO(NO_3)_3$	10—100	10%	4—6 X
$K_3Rh(NO_2)_6$	20—60	10%	3—5 X
$(NH_4)_3RuCl_6$	15—30	10%	3—5 X
$UO_2(C_2H_3O_2)_2$	25—75	5%	3—5 X
$RuCl_3$	—	—	

[a]) Growth is the estimate of the relative percentage of filamentous cells in a drop of culture fluid, and the relative length of the filaments is expressed in comparison to a control of normal size cells.

[b]) Undoubtedly contained $Pt(NH_3)_2Cl_2$.

studied the lysis, cell-breaking, of lysogenic bacteria by the same inorganic reagents (Table 2). The findings have been confirmed in several laboratories in different parts of the world. As a general word of warning we add that parallels between different biological effects do not imply identical sites or modes of action. We shall not be concerned therefore with the exact mode of attack of the reagents on the biological systems. Our concern will be to indicate in a general way *significant* chemical properties and ways in which the biochemical problems posed by these inorganic drugs can be tackled.

Table 2. *Lytic effects of platinum compounds on E.coli Tau⁻ strain (3)*

Compound	Oxidation State	Concentration Employed	
		25 μM	100 μM
Cis-Pt(NH₃)₂Cl₂	II	+	+
Trans-Pt(NH₃)₂Cl₂	II	—	—
PtenCl₂	II	(+)	+
Cis-Pt(NH₃)₂(NO₂)Cl	II	(+)	+
[Pt(NH₃)₄]Cl₂	II	—	—
Trans-Pt(NH₃)₂(NO₂)₂	II	—	—
K[Pt(NH₃)Cl₃]	II	—	(+)
Cis-Pt(NH₃)₂Cl₄	IV	O	O
Trans-Pt(NH₃)₂Cl₄	IV	O	O
PtenCl₄	IV	O	O
PtCl₄	IV	—	+
Pt(NH₃)₂(NO₂)₂Cl₂	IV	O	O
K₂PtI₆	IV	O	O
K₂PtCl₆	IV	O	O
Pt(en)₂Cl₂	IV	—	—
[Pt(NH₃)₂Cl₃]Cl	IV	—	O

+ = lysis;   O = inhibition;   — = no detectable effect.

In quite separate studies *Gillard* and co-workers (5) have examined the anti-bacterial activity of a number of additional inorganic compounds (Table 3). Table 3 should be compared with Table 2. Bacteriocidal action could arise in ways unrelated to the other biological activities mentioned above, see Table 4, so that selectivity between the two activities is not surprising. Finally *Dwyer* (6) investigated the action of a series of heavy transition metal compounds on nerve action and was able to show some drug-like properties. However as far as is known *Dwyer* did not test the activities of the platinum compounds on bacterial cells. These studies indicate a few areas in which inorganic drugs could be of value but a much wider testing of such drugs has been started.

Table 3. *Bacteriostatic action of trans-[RhL₄X₂]Y against staphylococcus aureus (5)*

Compound			Relative Activity
L	X	Y	
pyridine	Cl	Cl	1*
pyridine	Br	Br	4
4 *n*-propyl-pyridine	Cl	NO₃	30
4 *n*-propyl-pyridine	Br	NO₃	30
4-ethyl-pyridine	Br	Br	15 ᵃ)

Note: Corresponding Co(III) salts are inactive. The *cis*-salts, [Rh(II) B₂X₂]Y where B = *o*-phenanthroline, dipyridyl and ethylenediamine were inactive.

ᵃ) These compounds caused filamentous growth.

Table 4. *Effect of group VIIIb transition metal compoundsᵃ) on bacterial growth (concentrations of metal ions maintained for 2 hours at 8 p.p.m. in the continuous culture chamber) (2)*

A. Caused bacterial death	B. Caused no change	C. Caused elongation
CoCl₂	[Co(NH₃)₆]Cl₃	K⁺, NH₄⁺, H⁺—[PtCl₆]⁻ᵃ)
(NH₄)₂IrCl₆	K₂Ir(NO₂)₆	(NH₄)₂PtBr₆
NiCl₂	[Ni(NH₃)₆]Cl₂	(NH₄)₂PtI₆
(NH₄)₂OsCl₆		[Pt(en)₃]Cl₄
(NH₄)₂PdCl₄	*Cis* and *trans*	RhCl₃
	[Rh(en)₂Cl₂]NO₃	
[Rh(NH₃)₆Cl]Cl₂		(NH₄)₃RhCl₆
PdCl₂		[Ru(NH₃)₄ClOH]Cl
K₂PtCl₄	*trans*-Pt(NH₃)₂Cl₂	*cis*-Pt(NH₃)₂Cl₂
K₂PtCl₆	*trans*-Pt(NH₃)₂Cl₄	*cis*-Pt(NH₃)₂Cl₄

ᵃ) It is very important to note that the compounds added to bacteria are not necessarilly the active forms for many of the above compounds can hydrolyse or photo-aquate or react in other ways with ligands in the media.

It is also known that many B-group elements, Hg, Tl, Pb for example, are *poisonous*. In this article we shall not refer to them again until the concluding section as we shall be concerned to show potential beneficial effects of inorganic drugs.

II. The Chemistry

A. General Principles

We shall now describe the chemistry of those inorganic complexes which are known to have anti-tumour activity in an effort to outline the permutations which such molecules permit and to indicate possible functional modes. We start from the basic observation of *Rosenberg (1)*. *Cis* [PtCl₂ (NH₃)₂] is a very effective anti-tumour drug. Compounds related to it such as *trans* [PtCl₂(NH₃)₂] are ineffective. Out of a wide range of transition metal complexes tested few have proved to be effective. The successful compounds have certain common features which can be used to circumscribe some of the factors which are probably required for such a drug.

1. The complexes exchange only some of their ligands quickly. Such a restriction usually requires the central metal to be low-spin and means that some of the ligands bound in the complex will pass through biological membranes and into body compartments with the metal. This situation could be an extremely important source of analytical evidence when inquiring into the nature of the sites of action of the complexes. A radioactive tag can be included in the organic ligands of the metal.

2. The complexes are of one of three stereo-chemistries, see Fig. 1.

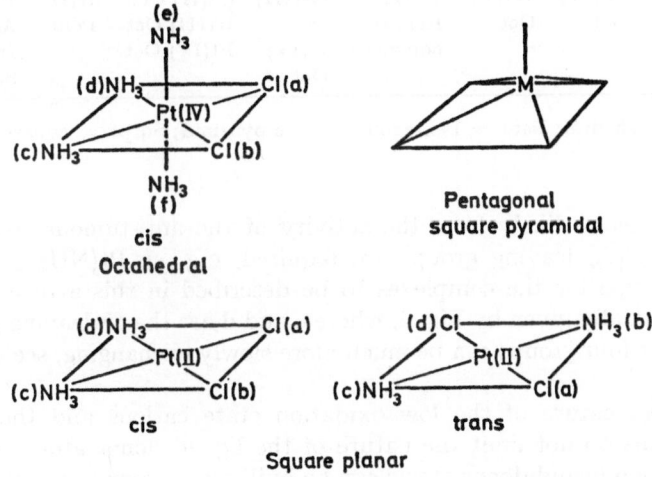

Fig. 1. The three stereochemistries of major importance for the majority of low-spin transition metal cations

5

Using rules (*1*) and (*2*), inspection of the Periodic Table limits the metal to the region given in Table 5. The ground state electronic configurations of the low-spin metals are included with a note of the stereochemistries adopted by any metal in a particular configuration. The oxidation states that are expected to be important in a biological system are also given. (The reactivities of the different oxidation states of a given metal are discussed later). The table emphasises that the low-spin configuration, d^8, leads, almost invariably, to a square-planar geometry, regardless of the oxidation state of the metal, that low-spin d^7 nearly always forms a pentagonal bi-pyramid, and that all other low-spin electronic configurations are usually octahedral (*7*).

Table 5. *Some metals and their oxidation states*

Electronic Configuration	d^1	d^2	d^3	d^5	d^6	d^7	d^9
First Row			Cr(III) Oct	Fe(III) Oct	Fe(II) Oct Co(III) Oct	Co(II) Pent & Oct	Ni(II) Oct & Sq. pl.
Second Row	Nb(IV) Oct	Mo(IV) Oct	Mo(III) Oct	Ru(III) Oct Rh(IV) Oct	Ru(II) Oct Rh(III) Oct Pd(IV) Oct		Pd(II) Rh(I) Sq. pl.
Third Row	Ta(IV) Oct	W(IV) Oct	W(III), Re(VI) polymeric	Os(III) Oct Ir(IV) Oct	Os(II) Oct Ir(III) Oct Pt(IV) Oct	Ir(II) Pent	Pt(II), Au(III) Ir(I) Sq. pl.

Oct = Octahedral; Pent = Pentagonal square pyramid; Sq. pl. = square plane

3. It seems likely from the activity of the anti-tumour complexes that two, *cis*, leaving groups are required, e.g. *cis*-Pt(NH$_3$)$_2$Cl$_2$. The formula-type for the complexes to be described in this article is then quite generally given by Fig. 2, where c and d are the *cis* leaving groups. The other four groups can be much more slowly exchanging, see note (1) above.

4. The nature of the low oxidation state cations and the above restrictions do not limit the nature of the ligand donor atoms greatly, Fig. 3. Such ligand donor atoms can be built into a very large number of ligands. Considering nitrogen as the donor atom, Fig. 4, shows some possible ligands.

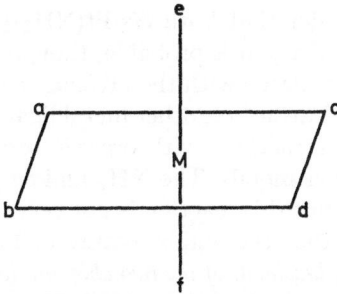

Fig. 2. The generalised set of ligand positions around a metal

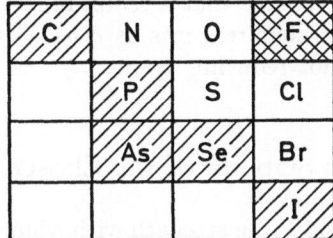

Fig. 3. Ligand donor atoms. The atoms in open blocks are the most common donors, those that are singly hatched are also common but are very improbable in a biological system, and the doubly hatched donor, F, is of very rare occurrence

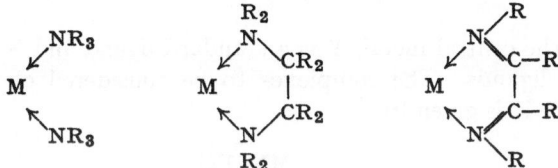

Fig. 4. The types of substitution which can be built into a ligand illustrated for nitrogen donors

Similar variations are possible for sulphur, carbon, phosphorus and arsenic, but not for halogen ligands of course. Mixed ligands such as 8-hydroxyquinoline can be utilised. Such a huge range of complexes are of potential value that we shall have to illustrate the relevant chemistry using general principles and a few examples. It should be noted that there is a very large number of coordination complexes which could act as drugs and that very few have been tested. This applies to drugs other than those of potential value against cancer.

5. When we consider that both cis-$Pt(NH_3)_2Cl_2$ and $Pt(en)Cl_2$ are active as anti-tumour drugs it is probable, though not absolutely certain, that the activity is associated with the *exchange reactions of the halide* and is not due to chemical attack on other metal ligands. The chlorine atoms cannot form bridged structures with organic residues though they can form bridges with other metals. The NH_3 and *en* groups could condense with carbonyl groups but this reaction is not very likely in platinum complexes. We conclude that the major feature of $Pt(NH_3)_2Cl_2$ is the ease of attack on Pt by *replacement of the two chloride ions*. Looked at through the eyes of an organic chemist cis-$Pt(NH_3)_2Cl_2$ is a bifunctional reagent with two *cis*-positions open to nucleophilic substitution. These positions are 3.3 Å apart.

6. It is highly improbable that a redox function is important in the activity of the anti-tumour reagents as cis-$Pt(NH_3)_2Cl_2$ is neither an interesting oxidizing nor reducing agent.

B. Thermodynamic Stability of the Chosen Complexes

In this section we consider the strength with which ligands are bound in the coordination sphere of the metals, i.e. the numerical value of the equilibrium constant, K, for the reaction

$$Y + -\overset{|}{\underset{|}{M}}-X \underset{\rightleftharpoons}{\overset{K}{\rightleftharpoons}} -\overset{|}{\underset{|}{M}}-Y + X$$

where M is the central metal, Y is a standard ligand and X is a variety of outgoing ligands. (The complexes to be considered obey the rules listed above). K is given by

$$K = \frac{[MX]\,[Y]}{[MY]\,[X]}$$

This is an important consideration in two ways for if a compound is to be useful it may be required that (1) it does not dissociate all its ligands but that (2) it is still able to exchange some of them. Fig. 5 shows that for a fixed oxidation state of a metal the stability of X, where X is

$$Cl^-, Br^-, I^-, RS^-, (PR_3),$$

relative to Y and where Y is

$$F^-, H_2O, RO^-, (or\ NH_3)$$

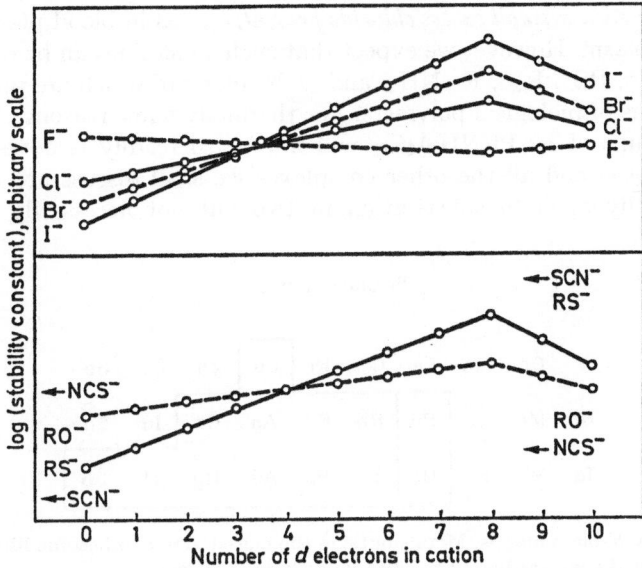

Fig. 5. The change in stability constant for different ligands with metals in the second and third transition series

increases as the metal gets heavier in a transition series. This generalization assumes that the other ligands, say *a*, *b*, *d*, *e* and *f* of the complex are fixed. In general for the metals under consideration thermodynamic stability, difficulty of replacement by water, follows the orders

$$I^- > Br^- > Cl^- \gg F^- \gg H_2O$$
$$RS^- > R_2S \gg NH_3 > H_2O$$

(carbon ligands $\gg$ phosphorus ligands $\gg$ halides)

(Inorganic chemists call such metal ions '*b*' class cations, Fig. 6). It does not follow that ligands rapidly replace one another in the above sequence, for this is a *kinetic* property (see next section) but ligands low in the order can never replace those higher in the order provided that the two free ligand concentrations are roughly equal. Other monodentate ligands which will bind reasonably strongly are OCN^-, SCN^-, N_3^-, while PR_3, CN^-, will bind very strongly. The importance of the orders lies in the ability of other ligands to exchange for group *c* (and *d*) in the initial complex. Thus if the groups *c* and *d* are weak they can exchange for several of the side-chains of proteins, saccharides, and nucleic acids. On the other hand we do not expect that complexes such as cis-Pt(NH₃)₂Cl₂ will lose chloride (here *c* or *d*) for water, $-CO_2^-$, $-SO_3^-$, alcohol-OH etc.

9

if there is even a small excess chloride present, e.g. as in blood plasma and cell cytoplasm. However we expect that such a chloride can be displaced by Br^-, I^-, RS^-, R_2S, $R-NH_2$, and $\geqslant N$, many of which are present as side chains of biological polymers. For thermodynamic reasons alone the NH_3 groups of *cis*-$Pt(NH_3)_2Cl_2$ will not be so readily removed. Thus $Pt(NH_3)_2Cl_2$ and all the other complexes we shall discuss are thermodynamically open to substitution in two but not necessarily in four positions.

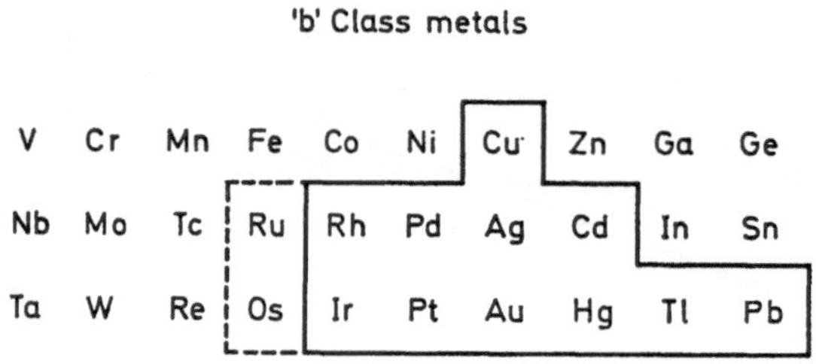

'b' Class metals

V	Cr	Mn	Fe	Co	Ni	Cu	Zn	Ga	Ge
Nb	Mo	Tc	Ru	Rh	Pd	Ag	Cd	In	Sn
Ta	W	Re	Os	Ir	Pt	Au	Hg	Tl	Pb

Fig. 6. The '*b*'-class metals. Metals outside the closed area can become like the '*b*'-class metals if they are bound to suitable ligands — see text

If a ligand can bind through two or more of its atoms it forms a chelate complex e.g

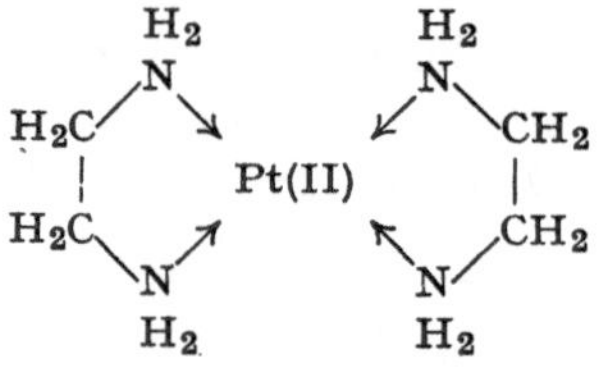

and such chelation greatly enhances stability. Thus positions in the coordination sphere can be blocked off from possible attack. Ethylene diamine, see Table 2, is very strongly bound by platinum.

If the incoming ligand X, which attacks the metal complex of Fig. 2, offers two coordinating groups, e.g. $^-S \cdot CH_2 \cdot CH_2 \cdot S^-$ or $H_2N \cdot CH_2 \cdot CH_2 \cdot NH_2$ then the simultaneous replacement of the two groups, *c* and *d*, can give very stable chelates of the type

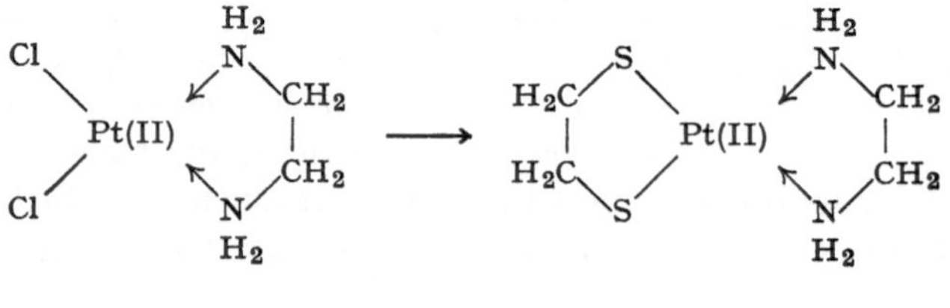

Such molecules may not be functional as drugs because they are *too stable*. However similar complexes are highly likely to be formed on reaction of the prescribed metal complexes with biological molecules which can easily act as chelating agents (see below). The fact that *cis*-Pt(NH₃)₂Cl₂ is active and *trans*-Pt(NH₃)₂Cl₂ is inert strongly indicates that the required reaction is one which leads to just such a chelated *cis*-structure with a protein, RNA or DNA. It is clear that, on thermodynamic grounds, such a chelated ligand will *not* be readily lost, once formed, from a metal. Thus if the complex M(abcdef) is looked upon as a reagent, positions can be blocked (thermodynamically) by making such complexes as

All such complexes can be thought of as potential „irreversible" labels of bio-molecules. A tris-chelate such as [Os(dipyridyl)₃]²⁺ will not behave as a reagent which has substitutional properties. Thus one limitation on the type of complex may well be that at least one or possibly two ligands should be monodentate as in [Pt(ethylenediamine)Cl₂].

The reagent *trans*-Pt(NH₃)₂Cl₂ has quite different potential reactivity for it could be a good but different bi-functional reagent. Organic chemists have observed that compounds such as bromo-acetone can react by condensation through the carbonyl group and by displacement through the bromine. A similar reaction is that of such di-alkylating agents as HN-(CH₂CH₂Br)₂, used in crosslinking DNA. Now the exact type of crosslinking through the bifunctional reagents is sensitive to the distance between the two reacting groups and their stereochemical demands. Thus both the isomers of Pt(NH₃)₂Cl₂ can cross-link but in very different ways.

The stability of binding of a group *c* or *d* is affected by the metal, its oxidation state, and the ligands both *cis* and *trans* to it. In practice it seems that the *trans* effect is usually considerably stronger than the *cis*. Thus if we look at the complex, Fig. 2, the ease of replacement of *c* and *d* by x and y, are affected not only by changes of M but also by *a* and *b* and *e* and *f* (*a* is *trans* to *c* and *b* is *cis* to *c* etc.). In general we can say that the better the donor strength of a *trans* (or *cis*) group the less stable is the binding of *c* and *d*. Ligands are easily placed in a rough order of *trans*-

11

weakening, influence (donor strength from equilibrium data) giving the orders of bonded atoms

$$C > N > O > F$$
$$\land \qquad \land \qquad \land$$
$$P > S > Cl$$

This means that the binding of a replaceable group c and d can be increased or decreased at will for the *cis* and *trans* groups a, b, e, f can be greatly altered. The series of complexes generated are comparable with a series of substituted benzoic acids or aniline bases in that here the ligand acceptor strength of M (for c or d substitution) can be adjusted by substitution on M while in organic chemistry the proton acceptor strength of $R-CO_2^-$ or $R-NH_2$ can be adjusted by substituting in R. Thus a whole family of potentially interesting complexes of graded property can be built even from *cis*-$Pt(NH_3)_2Cl_2$, by substituting the NH_3 groups, NH_2R etc., or by changing them to aromatic amine groups, or by changing them more radically to other donor atom centres. As in the design of all other drugs the choice of group is limited by considerations of solubility, toxicity, reactivity, etc. and compromises must be accepted, but a very wide range of possibilities exists.

C. Further Weak Coordination

Those complexes which have four or five of the possible coordination sites filled, see Fig. 1, can form weak bonds (*addition complexes*) in the fifth and sixth positions. It is known that $PtCl_4^{2-}$ will bind to *methionine*, *imidazole* and *amines* in such a fifth position, but very weakly. Such weak sites also offer initial positions of attack in an S_N2 substitution of say $PtCl_4^{2-}$, see below.

A further possible, but unproven, mode of bonding between *cis*-$Pt(NH_3)_2Cl_2$ and organic groups arises because of the high *dipole moment* of the *cis*-complex (compared with its absence in the *trans*-isomer) and the planar geometry of the molecule. Sandwich complexes, or charge transfer complexes, of the type formed between, say, hexamethyl benzene and tri-nitro-benzene or quinones are stabilized when one or both of the partners has a large dipole moment. Thus *cis*-$Pt(NH_3)_2Cl_2$ may bind weakly to planar electron-acceptor molecules like quinones. It could, however, be difficult to observe this interaction in solution where σ-interaction in the axial positions (discussed above) of the platinum complex will predominate over the sideways interaction. In a macromolecule where steric effects limit the geometry of a complex between *cis*-$Pt(NH_3)_2Cl_2$ and side groups the charge transfer complex may be formed. Alternatively the planar molecule may intercalate between bases of DNA. Dipole-dipole forces are known to be responsible in some cases for the observed intercalation of acridine dyes with DNA.

The importance of such weak coordination is difficult to establish but it will be discussed again later in connection with DNA and RNA.

D. The Oxidation State of the Metal

A major consideration before the ligand exchange equilibria can be considered with reference to biological systems is the stability of a particular oxidation state *in the biological medium*. Low-spin complexes undergo rapid one-electron oxidation and reduction. As a biological system operates at a low redox potential, say —0.5 to 0.0 volts, reduced, i.e. low valence, states of the metals are to be expected. The metal complexes, Ru, Os, Rh, Ir, Pd, Pt and Au should be reduced to the metallic state in fact but for the slow speed of this reduction. The metals of Fig. 6 will tend to go to the following redox states:

Cr(III), Co(III) or Co(II), Mo(V) or Mo(IV), W(V) or W(IV), Re(IV),

Os(II), Ru(II), Ir(III), Rh(III), Pt(II), Pd(II), Au(I)

whence the introduction of higher states, e.g. Pt(IV), should result in their quick reduction. As stated above Pt(IV) amine complexes may be the better compounds to add to the biological system for penetration but they are likely to act as Pt(II) complexes when inside cells, see Tables 1—4.

It is also important to see the effect of a simultaneous change in oxidation state and in a, b on the stability of the metal c, d system, Fig. 2. Very strong donors such as CH$_3^-$ joined to M in a higher oxidation state can make the *trans*-positions behave as they would in a low oxidation state e.g. in cobalt(III) complexes CH$_3$—Co(III) behaves very like Co(II). Thus

(CH$_3$)$_2$Pt(IV) (NH$_3$)$_2$Cl$_2$

could behave like

(NH$_3$)$_2$Pt(II)Cl$_2$.

The use of combinations of oxidation states with ligands of various donor power can then open the range of complexes which can behave like cis-Pt(NH$_3$)$_2$Cl$_2$ to a much wider group than is apparent from Table 5. Moreover the good donor, —CH$_3$, will not exchange and will therefore block positions. It is desirable to test many such organo-metallic complexes for activity. For example it is not generally recognised that a chemistry parallel to that of $[Pt(NH_3)_2Cl_2]$ can be induced in compounds of metals quite remote from the platinum group. *M. Green* and coworkers (Oxford) have shown that (cyclopentadiene)$_2MoCl_2$ will react with many amino-acids for example. Such compounds are now being tested for biological activity.

The relative stability of *cis* and *trans*-forms is important. The *trans* form is nearly always the more stable so that the rate of conversion is a limitation on the use of the *cis*-compound as a drug. This problem is overcome if two ligand positions are occupied by a chelating group for then only a *cis*-complex can be prepared.

E. The Kinetic Stability of the Chosen Complexes *(8, 9)*

The rate at which displaceable ligands leave or enter a complex is obviously important and is a quite separate consideration from the thermodynamic stability of the complexes. The reaction

$$-\overset{|}{\underset{|}{M}}-X + Y \xrightarrow{\ k\ } -\overset{|}{\underset{|}{M}}-Y + X$$

can have a half-life of very much less than a second to much greater than a month for the complexes under discussion. We can clarify the discussion of the rates in terms of

(i) the geometry of the complex, Fig. 1,

(ii) the metal and oxidation state involved

(iii) the ligands.

Under (i) the square and pyramidal complexes are often easier to substitute than the octahedral complexes for the obvious reason that they have open residual coordination sites, looking upon all the complexes as derived from an octahedron. The mechanism of substitution can then be the typical organic S_N2 attack. More usually the reactions of complex ions proceed by predissociation, S_N1, so that the important consideration is that c and d should be at least relatively good leaving groups.

In different oxidation states metals build *different geometries*, e.g. Pt(II) is planar and Pt(IV) is octahedral, and it is only to be expected that Pt(IV) will exchange ligands less readily than Pt(II). In general this is so. Thus on this basis Pt(IV) complexes would be expected to be more slowly 'acting' than Pt(II). It is not possible to make generalizations about the rate of ligand exchange on changing metal while keeping the geometry fixed, for these rates are very dependent upon the ligands and the exact number of d-electrons. For a given set of ligands however it has proved possible in a few cases to indicate some orders of magnitude. Thus square planar Au(III) complexes react approximately four orders of magnitude faster than those of Pt(II) although the nucleophilic properties of different reagents towards the Au(III) substrate parallel those of Pt(II). Again planar Pd(II) systems are about five orders of magnitude more reactive than the corresponding planar Pt(II) complexes.

Kinetic data have been most extensively examined for ligand exchange rates. The order of their rate of leaving (the order of 'good leaving groups' in organic chemistry) has been studied a number of times and we give a typical example.

The rates at which X leaves in a complex in which the other groups bound to the metal are fixed is shown by the series of reactions (10)

$$[Pt(dien)X]^+ + py \longrightarrow [Pt(dien)py]^{2+} + X^-$$

(dien) is $NH_2 \cdot CH_2 \cdot CH_2 \cdot NH \cdot CH_2 \cdot CH_2 \cdot NH_2$ and py is pyridine.

The rate of leaving follows the order

$$NO_3^- > H_2O > Cl^- > Br^- > I^- > N_3^- > SCN^- > NO_2^- > CN^-$$

which is probably just the stability order of the ligand binding to the metal. Thus one control of the attacking power (rate of attack) of a platinum compound rests in the choice of the leaving ligands.

Now the attacking group, in the above py, can also be changed, when it is found that heavier atoms tend to enter faster than lighter ones, suggesting an S_N2 mechanism. Substitution in *trans*-Ptpy₂Cl₂ by nucleophiles gives rise to the following relative rates of attack (11)

$$S_2O_3^{2-} > SC(NH_2)_2 > PhS^- > SeCN^- > SO_3^{2-} > I^- > Br^- >$$

$$N_2H_4 > N_3^- > NO_2^- > NH_3 > Cl^- > CH_3O^-.$$

The order is roughly

$$RS^- > I^- > Br^- > -N > Cl^- > RO^-.$$

It is generally considered that such substitution reactions go through a trigonal bipyramid intermediate and it is the polarisability of I⁻ and sulphur ligands which makes them such good attacking groups. The fact that they are good entering but not such good leaving groups is reflected too in the high thermodynamic stability of their complexes described above. Pt(NH₃)₂Cl₂ should be rapidly attacked by methionine and a stable complex should be formed.

The rate of leaving of a group is also dependent upon the other groups present and in particular the group *trans* to it (*12*). Table 6 gives some data. Whereas the thermodynamic data establish a stability order for binding of ligands

$$CN^- > OH^- > NH_3 > SCN^- > I^- > Br^- > Cl^- \gg F^- \simeq H_2O$$

The kinetic *trans* effect gives the order

$$CN^-, C_2H_4 > SC(HN_2)_2, PR_3, SR_2 > NO_2^- > I^-, SCN^- > Br^- > Cl^- >$$
$$NH_3, py, RNH_2 > OH^- > H_2O$$

Table 6. *Trans-effect on the rates of reaction of some Pt(II) complexes with Pyridine (12)*

$$\begin{array}{c}
L \\
\diagdown \\
Pt \\
\diagup \\
Et_3P
\end{array}\!\!\!\begin{array}{c}
PEt_3 \\
\diagup \\
 \\
\diagdown \\
Cl
\end{array} + Py \longrightarrow \begin{array}{c}
L \\
\diagdown \\
Pt \\
\diagup \\
Et_3P
\end{array}\!\!\!\begin{array}{c}
PEt_3 \\
\diagup \\
 \\
\diagdown \\
Py
\end{array} + Cl^-$$

L	k_1 (min^{-1})
H	1.1
PEt$_3$	1
CH$_3$	$1 \cdot 10^{-2}$
C$_6$H$_5$	$2 \cdot 10^{-3}$
Biphenyl	$1 \cdot 10^{-3}$
Cl	$6 \cdot 10^{-5}$

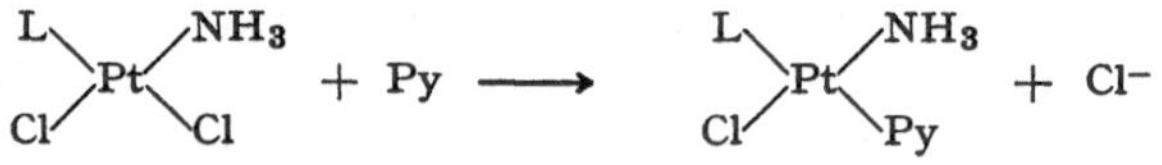

$$\begin{array}{c}
L \\
\diagdown \\
Pt \\
\diagup \\
Cl
\end{array}\!\!\!\begin{array}{c}
NH_3 \\
\diagup \\
 \\
\diagdown \\
Cl
\end{array} + Py \longrightarrow \begin{array}{c}
L \\
\diagdown \\
Pt \\
\diagup \\
Cl
\end{array}\!\!\!\begin{array}{c}
NH_3 \\
\diagup \\
 \\
\diagdown \\
Py
\end{array} + Cl^-$$

L	k_1 (sec^{-1}M^{-1})
Cl$^-$	6.3
Br$^-$	18
NO$_2^-$	56
C$_2$H$_4$	Very fast

The big difference in these series is the relative position of NH$_3$ and OH$^-$, ligands from the first row of the Periodic Table. These ligands behave as atoms with a high charge density and low polarisability so that their affinity for protons also greatly exceeds that of most of the other ligands. The *trans*-effect is more dependent on polarisability than is the stability. Whatever the origin of the differences in order these differences can be used in the design of potential drugs. Whereas the groups NH_3 and PR_3 may be equally effective in preventing reaction at a given coordination site of platinum, once the complex is bound to a poor leaving group e.g.

$$\begin{array}{c} H_3N \\ \end{array} \!\! > \!\! Pt \!\! < \!\! \begin{array}{c} N\text{---protein} \\ N\text{---protein} \end{array}$$

then the low *trans*-effect of NH_3 will make the reaction relatively irreversible. The formation of

$$\begin{array}{c} R_3P \\ \end{array} \!\! > \!\! Pt \!\! < \!\! \begin{array}{c} N\text{---protein} \\ N\text{---protein} \end{array}$$

could be much more reversible. A more detailed discussion is not fruitful here but it should be noticed that the whole basis of the synthesis of various Pt(II) compounds in inorganic chemistry has been the qualitative and sometimes quantitative utilisation of the competing stability and kinetic factors. The same principles would apply to drug action and the differentiating features inherent in the rate and stability data could lead to a fine control of activity.

F. Parallels between Pt(abc)⁺ and C(abc)⁺

The chemistry of a carbon(+) centre is best looked at through the chemistry of carbon in a ketone or an aldehyde which we can write

$$\begin{array}{c} R_1 \\ \end{array} \!\! > \!\! C^+ \!\! - \!\! O^-$$
$$R_2$$

The types of reagent which attack such groups *rapidly* and attach (bind) themselves strongly are $-SO_3H^-$ (through S), CN^- (through C), NH_3 (through N), but H_2O does not add so readily. In aldol condensations carbon-carbon bonds are formed. Thus carbon(+) reacts somewhat like [Pt(abc)]⁺ in that it takes on carbon, sulphur and nitrogen readily. The attack of carbonium ions on biological molecules is well understood and could assist in the development of work on Pt(abcd), see below.

G. Comparison of Thermodynamic and Kinetic Properties

As pointed out by *Grinberg* and *Borzokova* (13) it can happen that the lability of groups (rate of leaving) can be inversely related to the stability of the complexes if the group *trans* to the leaving group is not held con-

A. J. Thomson, R. J. P. Williams, and S. Reslova

stant. In the case of the complexes PtX_4^{2-} where X is CN^- or a halide the order of thermodynamic stability is

$$CN^- > I^- > Br^- > Cl^-$$

and the rate of exchange is greatest in the same order, see Table 7. Very strongly bound ligands can produce a very strong labilising effect on other very strongly bound ligands. Thus if it is desirable to produce a compound in which some groups are strongly held and not labile and others are not strongly held and labile this can be brought about by a wide but careful choice of different ligand systems. The absence of parallel between thermodynamic data and kinetic data is also apparent in some organic reactions where pK_a, the thermodynamic acid dissociation constant, may not be related to the rate at which a group attacks or is attacked, nucleophilicity. In the transition metal series of compounds both the thermodynamic stability and the nucleophilicity are known and although it may not be possible to understand the theoretical reasons for their behaviour it is possible to choose reagents which show a thermodynamic or a kinetic discrimination of high selectivity. In principle then there is a great potential source of drug molecules amongst metal complexes and it now remains to be seen to what uses such compounds can be put. They have the advantage over organic molecules that they are easy to follow in a biological system as we shall now show.

Table 7. *Comparison between the rate of exchange in PtX_4^{2-} and the equilibrium constant, K, for $Pt^{2+} + 4 X^- \rightleftharpoons PtX_4^{2-}$*

X^-	CN^-	I^-	Br^-	Cl^-	NH_3
Half-life	1	5	8	280	273 days
log. K.	44	29.6	20.5	16.6	35.3

III. Physico-Chemical Properties and their Probe Potential

A. General Considerations

In this section we outline briefly the spectral and magnetic properties of complexes of the metals of Table 5. These are the properties that enable the stereo-chemical and electronic structures of metal complexes to be determined in solution and, hence *in a biological environment*. A study of these properties will be necessary to understand the nature of the interaction of the anti-tumour compounds with biological systems.

18

Thus we shall be concerned with properties that furnish information about the nature of the ligands, the oxidation state of the metal, and the geometry of the field of ligands. Techniques such as radio-isotope tracer studies, neutron-activation analysis, and electron microscopy are powerful methods for locating a metal within constituents of the cell and are particularly suited to heavy-metal rather than organic drugs but since they do not provide information about the chemical environment of the metal they will not concern us here. After each section below we shall give an example, not necessarily from platinum chemistry, where the method has been used with success in biochemistry.

B. Electronic Absorption Spectra

The electronic absorption spectrum of a solution of $[PtCl_4]^{2-}$ ions, see Fig. 7, can be divided into two regions; one consisting of absorption bands of extinction coefficients less than 500 l · mol⁻¹ at energies below 35,000 cm⁻¹, and a region containing intense bands, extinction coefficients greater than 10,000 l.mol⁻¹ and usually at higher energy *(15)*. The weak absorption bands arise from transitions of the electrons of the unfilled *d*-shell (*d—d transition*) and although formally forbidden, gain intensity through vibrational perturbations and through borrowing from the intense allowed bands. This intensity borrowing becomes more pronounced the lower the symmetry of the complex ion. The intense bands arise from transitions of electrons in orbitals predominantly on the ligand to orbitals primarily on the metal, and are termed *charge-transfer* (C. T.) *bands.* in this case, *ligand-to-metal* (*L → M*) *bands (16)*.

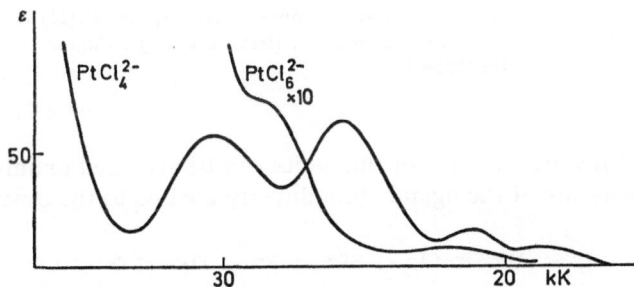

Fig. 7. Electronic absorption spectra, *d—d* region, of Pt(II)Cl₄²⁻ and Pt(IV)Cl₆²⁻ in aqueous solution (from *Jorgensen, C. K.*: Absorption Spectra and Chemical Bonding in Complexes. London: Pergamon 1962)

(1) *d—d bands*. With successive replacement of the chloride ions of $PtCl_4^{2-}$ by ammonia ligands the *d—d* bands move progressively to higher energy, see Fig. 8. Ligands can be arranged in an order, the spectrochemical series, empirically determined and now with a firm theoretical basis, so that for a given metal ion ligands high in the series bring about the largest shifts of the metal *d—d* bands towards high energy.

$$I^- < Br^- < Cl^- < R_2S < (M{\leftarrow})SCN^- < F^- \simeq urea \simeq OH^- \simeq N\overset{O^-}{\underset{O}{\diagup\diagdown}}$$

$$< [oxalate]^{2-} < H_2O < SCN^-(\rightarrow M) < glycinate < pyridine < NH_3 <$$

$$ethylenediamine \ll CN^- \simeq CO$$

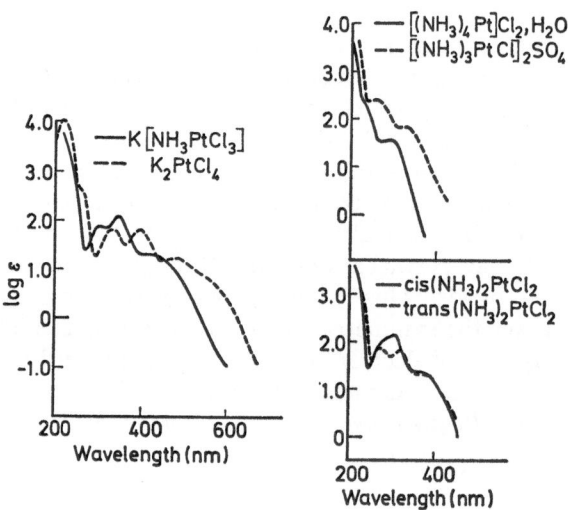

Fig. 8. Electronic absorption spectra of some planar square Pt(II) complexes of chloride and ammonia in aqueous solution (from *Chatt, J., Gamlen, G. A., Orgel, L. E.*: J. Chem. Soc. 486 (1958))

An abbreviated version of this series can be arranged mainly according to the nature of the ligand atom directly bonded to the central metal ion:

$$I^- < Br^- < Cl^- < S^{2-} < F^- < O^{2-} < N < C$$

A complex containing a mixture of ligands will have a *d—d* spectrum at energies close to the mean of the spectra of the complexes with only one type of ligand (*17*).

There is also a dependence of the energy of the $d-d$ spectrum on the nature and the oxidation state of the metal. For a given type of ligand the energies of the $d-d$ bands will lie at higher energies the higher the oxidation state of the metal. Fig. 7 compares the $d-d$ spectrum of $[Pt(II)Cl_4]^{2-}$ with $[Pt(IV)Cl_6]^{2-}$ (18). Again, for a given ligand, metals from the 3rd transition series display $d-d$ bands at higher energies than those from the 2nd transition series, and from the first transition series. Thus we may state a spectrochemical series of central metal ions and for metal oxidation states for the same ligand.

Co(II) < Fe(II) < Fe(III) < Cr(III) < Rh(III) < Pd(IV) < Ir(III) < Pt(IV)

A further useful characteristic of the $d-d$ spectrum is revealed by examination of the spectrum of cis-Pt(NH₃)₂Cl₂ and trans-Pt(NH₃)₂Cl₂, Fig. 8. The splitting between the prominent bands is rather different, and the intensity of the cis spectrum is markedly higher. This illustrates two further general rules; that the splitting of $d-d$ spectrum is strongly dependent upon the symmetry of the field of ligands and that the lower symmetry complex, in this case the cis-isomer, has the more intense spectrum. We see therefore that a study of the $d-d$ spectrum could supply a great deal of information about the nature of the ligands and the geometry of a complex ion.

The only potential ligand present in a biological medium which is close to Cl in the spectro-chemical series is sulphur (cf. the spectrum of trans-Pt(II) (NH₃)₂(SEt₂)₂ and trans-Pt(II) (NH₃)₂Cl₂ (19). All other ligands are rather high in the series. Therefore the replacement of chloride ions of cis-Pt(NH₃)₂Cl₂ by ligands from biological materials will result in a shift of the d-d spectrum to high energy and, in most cases, this will cause the spectrum to overlap badly with the absorption spectrum of proteins and nucleic acids. This emphasizes the need to find analogues of the platinum anti-tumour compounds containing metals from the first transition series. Compounds such as

cis-[Cr(III) (NH₃)₄Cl₂]⁺   and   cis-[Co(III) (NH₃)₄Cl₂]⁺

should be tested for activity since much of their $d-d$ spectrum is well below 30,000 cm⁻¹.

2. *Charge transfer bands*. The charge-transfer bands which characteristically have intensities 10^4 l.mol⁻¹ lie at energies determined mainly by the oxidizing power of the metal relative to the reducing power of the ligand if the transition is L → M, and, vice versa for M → L transitions. A correlation between redox properties and transition energies is expected if the transition corresponds to the removal of an electron from a ligand

A. J. Thomson, R. J. P. Williams, and S. Reslova

and the placing of it on the central metal ion. This is illustrated clearly
in Fig. 9 which shows the spectrum of $[Co(NH_3)_5X]^{2+}$ where X^- is
NH_3, F^-, Cl^-, Br^- or I^-. All the high intensity bands in these spectra are
$L \rightarrow M$ charge-transfer. As the reducing power of the ligand increases from
Cl^- to Br^- the energy of the bands drop (20), or, with constant ligand,
as the oxidizing power of the metal rises from Ir(III) to Ir(IV) the band
system also moves to lower energy. (The increased detail seen in some
of the bands of the bromo-salts arises from the large atomic weight of
the latter ion compared to chloride and results from spin-orbit coupling).

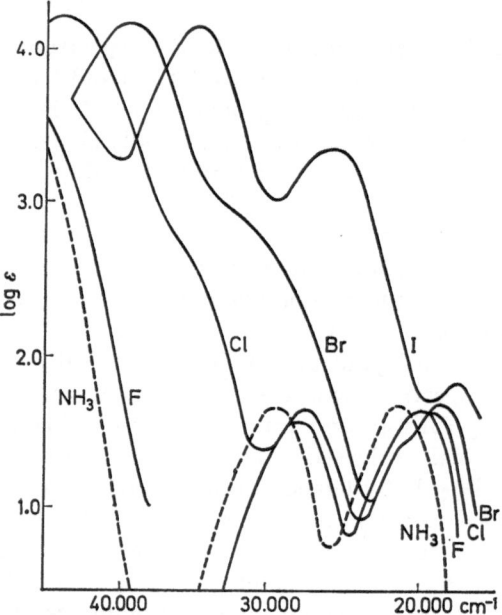

Fig. 9. Electronic absorption spectra in aqueous solution of $Co(NH_3)_5X$ where
X is NH_3, F^-, Cl^-, Br^-, I^- (from *Linhard* and *Weigl*: Z. Anorg. Allgem. Chem. *226*,
49 (1951))

The intensity of charge transfer bands makes them potentially
valuable for following dilute solutions of metal complexes in their interaction with biological molecules. One of the main obstacles to the study
of the interaction of, say, *cis*-$Pt(NH_3)_2Cl_2$ with proteins by spectro-
photometry arises from the low intensity of its absorption as well as
the overlap with the absorption bands of the protein. By using a more
reducing ligand, such as Br^-, instead of Cl^-, $L \rightarrow M$ charge-transfer bands

can be brought to lower energy. Thus complexes such as *cis* and *trans*-Pt(IV) $(NH_3)_2Br_4$, provided they are biologically interesting, would be easier to study in their interactions with biological ligands. The *cis*- and *trans*-isomers of $[Ru(III) (NH_3)_4Cl_2]^+$, orange and red-orange, respectively, should be examined as should *cis*- and *trans*-$[Ru(III) (NH_3)_4Br_2]^+$, red and purple respectively.

Studies of the absorption spectra of the anti-tumour compounds and the spectra of their complexes with biological molecules will undoubtedly be the most convenient and effective way of examining the possible interactions, of measuring binding constants, and of determining rate constants for complexation.

As examples, Table 8 records some observations on $d-d$ and charge transfer absorption bands in metal/protein systems. The examination of the spectrum of cobalt carbonic anhydrase ($d-d$) and of iron conalbumin (charge-transfer) permitted a prediction of the ligands from the protein to the metal. The predictions have now been substantiated by other methods.

Table 8. *Absorption bands in metalloproteins*

Protein	$d-d$ Band (nm)	Charge-transfer Band (nm)
Co(II) Carboxypeptidase	530 (150)	—
Co(II) Alcoholdehydrogenase	620 (800)	350 (3,000)
Co(II) Carbonic anhydrase	555 (400)	—
Fe(III) Conalbumin	—	470 (3,300)
Fe(III) Hemerythrin	750 (100)	330 (3,200)
Fe(II) Rubredoxin	2,000 (50)	320 (5,000)

Molar extinction coefficient is in brackets.

C. Further Techniques Related to Absorption Spectroscopy

(1) *Magnetic circular dichroism (MCD)*. This technique provides information about the geometry of a field of ligands bound to a central metal (21). The method consists of the measurement of the difference in the absorption of right and left circularly polarized light in the presence of a strong magnetic field co-linear with the light path through the sample. Measurements are now conveniently carried out on solution samples at room temperature in the bore of a superconducting solenoid which has been lowered into the cell compartment of a standard circular dichroism spectrometer. The technique is a method of observing Zeeman splittings

of electronic energy levels when the splitting due to the magnetic field is much smaller than the width of the absorption bands. The MCD spectrum of the $PtCl_4^{2-}$ ion was most useful in verifying the assignment of its spectrum (27). The MCD spectra of cis- and trans-Pt(II) $(NH_3)_2Cl_2$ and particularly the cis- and trans-Pt(IV) $(NH_3)_2Cl_4$ isomers are expected to be very different owing to their different symmetries and should distinguish the isomers more effectively than their absorption spectra do. These measurements are in hand.

(2) *Natural circular dichroism (optical activity)*. Although circular dichroism spectra are most difficult to interpret in terms of electronic structure and stereochemistry, they are so very sensitive to perturbations from the environment that they have provided useful ways of detecting changes in biopolymers and in complexes particularly those remote from the first co-ordination sphere of metal complexes, that are not readily apparent in the absorption spectrum (22). It is useful to distinguish between two origins of the rotational strength of absorption bands.

All chromophores gain rotational strength, that is, their absorption bands will show a difference in the absorption of right and left circularly polarized light, when they are dissolved in an optically active medium. For example, the absorption bands of $[PtCl_4]^{2-}$ become optically active when the ion is dissolved in an optically active alcohol. If this ion is bound, without loss of ligands, to a protein some of the bands will undoubtedly gain rotational strength. Similarly, if this planar ion were slipped between the base pairs of DNA (intercalated) the $d-d$ bands would gain intense optical activity. The dyes such as proflavine and acridine orange which are normally optically inactive become intensely active when bound either to the phosphate backbone of the DNA or when intercalated (23). Thus the technique of circular dichroism could become a powerful method of detecting weak interactions of this kind with biopolymers.

Optical activity in metal complexes may also arise either if one of the ligands bound to the metal in the first co-ordination sphere is itself optically active or if the complex as a whole lacks a centre of inversion and a plane of symmetry. Thus all octahedral cis-complexes of the tris- or bis-chelate type have two isomeric forms related by a mirror plane, the d- and l-forms. These species have circular dichroism spectra of identical intensities but opposite in sign. The bands in the circular dichroism spectrum are, of course, modified if ligand exchange occurs but they are also exceedingly sensitive to the environment beyond the first co-ordination sphere. This effect has been used to obtain association constants for ion-pair formation. There also exists the possibility that, if such compounds display anti-tumour activity, only one of the mirror isomers will be effective.

Table 9 gives some cases where the rotational strengths of absorption bands have been measured in metalloproteins. At the present time these changes are not used to diagnose the nature of the ligands of the metal but rather they have been used to follow minor changes at the metal when substrates or inhibitors interact with the metals. The sensitivity of CD and MCD measurements to very small changes in the metal environment make them very attractive for protein/metal complex studies.

Table 9. *Rotational strengths of absorption bands of metalloproteins*

Protein	Absorption Maximum (nm)	Rotation Amplitude per mole
Fe(III) Adrenodoxin	480	900°
Fe(III) Hemerythrin	500	70°
Fe(III) Conalbumin	470	20°
Cd(II) Metallothionine	268	1570°

D. Electron Spin Resonance (ESR)

If a molecule contains one or more unpaired electrons it is usually possible to detect an electron spin resonance signal and at a very low concentration of unpaired electrons, commonly 10^{13} spins with modern instruments. Several pieces of information can be obtained in this way. The number of unpaired spins can be found, the symmetry of the molecule in the region of the unpaired electron can be determined, and, if the unpaired electron is delocalized over nuclei with nuclear spins, then the extent of delocalization can be determined. Perhaps more importantly for our purpose, the rotational time of molecules can be determined from line shape studies.

Recently *McConnell* has introduced the technique of labelling proteins lipid membranes, and DNA, with a *stable organic free radical*, the nitroxide radical, see reference (24). A measurement of the line shapes of the ESR signal of this label has revealed the rotational times of the molecules to which it is bound. If paramagnetic analogues of the anti-tumour compounds are found then their progress within a cell will be readily followed using ESR. Complexes of all the metals of Table 1 with odd numbers of electrons (d^1, d^3, d^5 and d^7) are potentially useful. Thus

cis-[Cr(III) (NH$_3$)$_2$Cl$_4$] (d^3),

cis-[Ir(IV) (NH$_3$)$_2$Cl$_4$] (d^5), and

cis-[Ru(III) (NH$_3$)$_2$Cl$_4$] (d^5)

A. J. Thomson, R. J. P. Williams, and S. Reslova

are some examples. Indeed, it has already been observed by *Gale, Howle,* and *Smith* (*25*), that a neutral complex formed by the irradiation of $[Ir(IV)Cl_6]^{2-}$ in ammonia buffer is effective in forcing filamentous growth of *E. coli*. Provided that photo-reduction to the Ir(III) state has not occurred (see below) and provided also that reduction does not occur within the cell it would appear that a spin labelled complex is available.

Failing this approach, it should be possible to attach *McConnell's* type of spin label to a ligand such as ethylene diamine and to carry it into a cell with Pt(II).

In Table 10 there are some examples of EPR signals obtained from metallo-proteins. Most of this work has been conducted at low temperatures and it is now necessary to develop probes which contain metals and which will give signals at room temperature, e.g. d^1 complexes. Such a label could be of use in many explorations of activity in biological systems.

Table 10. *Electron spin resonance signals of metalloproteins*

Protein	g-Values
Cu(II) Erythrocuprein	2.05, 2.35
Cu(II) Plastocyanin	2.05, 2.35
Fe(III) Conalbumin	4.25, 8.8
Fe(III) Ferrodoxin	'1.94'
Mo(V) Xanthine oxidase	1.95, 1.97

E. Nuclear Magnetic Resonance (NMR)

NMR techniques can be useful for the determination of the site of action of a metal complex, of its binding constant, and rate data, particularly the rate of exchange of ligands. There are two types of experiments that can be performed.

In the first case, an attempt is made to *detect the characteristic resonance of a given nucleus*, most usually the proton, but since they are found in such low abundance in most chemical and all biological systems, the resonance of other nuclei such as [195]Pt, ^{31}P, [35]Cl could be studied (*26*). Nuclei such as ^{19}F could be substituted into ligands of the metal complexes. The shift of the resonance from some standard value (the chemical shift) gives an indication of the chemical environment of the resonating nucleus. But this general method suffers from the severe disadvantage, lack of sensitivity, particularly for nuclei other than protons. For example, [195]Pt resonances are only detected at about 1 M in platinum.

26

The second method is likely to prove the more useful. If a paramagnetic center of *fast* electronic relaxation time is placed close to a resonating nucleus the *chemical shift of that nucleus is increased enormously.* Many of the paramagnetic complexes suggested in the last section would shift the resonances of protons adjacent to them. In this way it is possible to observe the resonances of protons in the immediate vicinity of a paramagnetic ion in a protein containing many thousands of protons (27). Elegant examples of such studies are the use of the Co^{2+} (d^7) (27) and Ln^{3+} ($4f^n$) ions bound to lysozyme (28). It is not always necessary that the perturbing metal should have an unpaired electron. Many metal complexes have a small paramagnetism that is independent of temperature (TIP). This arises from the contribution to the ground electronic state of paramagnetic excited states and becomes larger the closer the excited states approach the ground state. Pt(II) and Co(III) complexes are well known for their appreciable TIP (29).

Rates of exchange processes, particularly *ligand exchanges*, can be determined from measurements of the shapes of NMR lines. This is the analogue of the determination of rotational times of complexes from the shapes of ESR lines. Provided that the nucleus being observed is exchanging magnetic environments fast compared with the lifetime of a nuclear resonance transition, the magnetic field observed by the nucleus, which is the sum of the applied field and the field due to the magnetic environment, is constant and the resonance is sharp, the line-width narrow. But when the exchange process becomes comparable in time with a nuclear resonant transition the line will be broadened. Typical lifetimes studied are of the order of 10^{-1} to 10^{-3} cps (30). Fe^{3+} (d^5) bound to conalbumin is an example (31).

The presence of the chloride ion in *cis*-[Pt(II)(NH₃)₂Cl₂] suggests the use of the ^{35}Cl resonance in experiments of the type carried out by *Bryant* and *Baldeschwieler* in examining the environment of Cl ions in DNA. (32, 33).

F. Vibrational Spectra

The vibrational spectrum of a metal complex is one of the most convenient and unambigious methods of characterization. However, it has not been possible to study the interactions of metal ions and biological polymers in this way since the number of vibrational bands from the polymer obscure the metal spectrum. The use of laser techniques for Raman spectroscopy now make it very likely that the *Raman spectra* of metals in the presence of large amounts of biological material will be measured (34). The intensity of Raman lines from metal-ligand vibrations can be

A. J. Thomson, R. J. P. Williams, and S. Reslova

very high since these are the vibrations displaying large polarizability changes. Recent work has suggested the use of resonance enhancement phenomena to observe the Raman lines of chromophores on proteins (35).

The *far infra-red spectra* of many Pt(II) and Pt(IV) complexes are known. For example, in the infra-red spectrum of the *cis*- and *trans*-complexes of Pt(II)L$_2$Cl$_2$ and Pt(IV)L$_2$Cl$_4$ the band due to the Pt—Cl stretch is very strong. In the *trans* complexes the Pt—Cl stretch is at 339.5 ± 3.5 cm^{-1} for Pt(II) and at $\sim$350 cm^{-1} for Pt(IV) (36, 37). The frequency is almost independent of the ligand L, as expected, except that for *trans*-Pt(NH$_3$)$_2$Cl$_2$ the frequency is at 331.5 cm^{-1}. It has been suggested that H-bonding in the crystal interferes in this case. For the *cis*-complexes, the Pt—Cl stretching frequency depends upon the ligand, L, *trans* to the chloride ligand. A few examples are quoted

L	ν (cm^{-1})
(Et$_2$S)$_2$	330.0 (mean)
(py)$_2$	327.9 (mean)
(NH$_3$)$_2$	326.0 (mean)
(Et$_3$P)$_2$	303.2 (280.2)

Two lines are observed in the spectrum of the *cis*-complex (the mean of the two is given in the table) whereas only one is found in the *trans*-form. This arises from the lower symmetry of the *cis*-form.

The vibrational spectrum of *cis*- and *trans*-[Pt(II) (NH$_3$)$_2$Cl$_2$] in the crystalline phase has also been measured with laser-Raman techniques by Dr. *Hoeschele* of the Biophysics Department, Michigan State University. This technique shows great promise.

IV. Photochemistry of Metal Complexes

The photo-sensitive nature of metal complexes has played an important role in the admittedly fortuitous generation of species active in forcing filamentous growth of *E. coli*, and it is well to be aware of these effects when metal complexes are incubated in biological media in light. Photochemical effects may also be important in the interpretation of the results of the heavy atom labelling of protein crystals since the metal complexes are left to react for long periods in the mother liquor and we cannot assume that this process has always been conducted in the dark. Since the absorption of a photon of light adds free energy to a chemical system the thermodynamic and kinetic arguments given earlier can be invalidated

28

if the process is a photo-chemical one. Three types of photoreaction are observed:

a) photo-ligand substitution, usually photo-solvation,

b) photo-reduction or oxidation,

c) photo-isomerism (*38*).

All members of the family $[Pt(NH_3)_nCl_{4-n}]^{n-2}$ photo-aquate (*39*). *Cis*-$[Pt(NH_3)_2Cl_2]$ yields the product $[Pt(NH_3)_2(H_2O)Cl]^+$ with a quantum yield 0.46 at 350 nm and 25 °C (*40*). The yield is dependent upon both wavelength and temperature. The *trans* isomer also photo-aquates but with a quantum yield of 0.01. There has been a report that with light below 200 nm photo-oxidation can take place generating polynuclear Pt(II)—Pt(IV) species (*41*). $[Pt(NH_3)_3Cl]^+$ photolyses to give ammonia and chloride aquation (*42*). $[Pt(IV)Cl_6]^{2-}$ photo-aquates.

Photo-isomerism is also found in this group of compounds. *Cis*-$Pt(NH_3)_2(H_2O)_2^{2+}$ photo-isomerizes to *trans* with a quantum yield of about 0.1 at 363 nm (*40*). Photoisomerism of Pt(glycine)₂ from *cis* to *trans*, but not its reverse is also reported.

Photo-oxidation or reduction is often found if the complex is irradiated in the charge-transfer bands (see above); photo-oxidation of the metal occurring if the transition is M →L. Thus the photochemical generation from $Ir(IV)Cl_6^{2-}$ of a species active in forcing filaments of *E.coli* may well involve the photoreduction of Ir(IV) to Ir(III) since the intense bands in the visible spectrum of $Ir(IV)Cl_6^{2-}$ are L →M charge-transfer bands. A report has appeared of the photo-aquation of $IrCl_6^{2-}$ (*43*).

V. Interaction of Metal Complexes with Biological Ligands and Macro-molecules

A. Introduction

In this section we summarise the manner in which '*b*'-metals, Fig. 6, and where possible specifically the platinum complexes of concern here, interact with biological molecules. Some radio-tracer studies have been carried out on the distribution of platinum complexes in whole bacteria grown in media inocculated with the metal ion. The results are summarised in Table 11. It is noteworthy that the bacteriocidal complex $[PtCl_6]^{2-}$ was taken up almost entirely by the cytoplasmic protein whereas the filamentous forming neutral species, $[Pt(NH_3)_2Cl_4]$, was also bound to the nucleic acid fraction and to metabolic intermediates which comprise amino-acids, peptides, and small nucleic acid molecules. The lipid material contained surprisingly little of any of the species. The distribution of the metals over a wide variety of chemical molecules is clearly to be expected in all whole organisms and the task of finding the functionally significant interaction is thus made the more difficult. In the following sections then we shall treat the components of biological systems in turn, looking at a wide variety of biological ligands and metals. Only at the very end of the article shall we return to the problem of whole biological systems.

Table 11. *Average distribution of* ^{191}Pt *in bacteria* (44)

Fraction	Percentage of Radioactivity			
	Neutral Species[a])			$PtCl_6^{2-}$
	B.cereus	S.aureus	E.coli	E.coli
Metabolic intermediates	74	60	19	1
Lipids	2	1	6	3
Nucleic acids	19	20	30	1
Cytoplasmic protein	5	19	45	96

[a]) Platinum complex generated photochemically from $PtCl_6^{2-}$ in ammonia buffer and known to contain primarily $Pt(NH_3)_2 Cl_4$ although the ratio of cis to trans isomer is unknown.

B. The Reaction of Platinum Complexes with Amino-acids

The interaction of planar Pt(II) complexes with aminoacids have been studies mainly by Russian workers particularly from the group of *Volsteyn*. It was found that $[PtCl_4]^{2-}$ and *cis*- and *trans*-$[Pt(NH_3)_2Cl_2]$ can be substituted at the chloride positions by the following amino-acid groups: $-NH_2$, $-CO_2^-$, $-SCH_3$, and the ring nitrogens of imidazole (histidine). No reports have appeared of interactions with aromatic groups such as phenolate (of tyrosine), ring nitrogen of tryptophan or with sulphydryl groups of cysteine although all of these are potential coordinating groups. The rules given earlier for the thermodynamic and kinetic control of the reactions of Pt(II) complexes with simpler ligands clearly apply to the attack by the amino-acids as is recognised by the Russian workers.

The range of complexes which can be formed by just one amino-acid, glycine, and one Pt(II) complex, $PtCl_4^{2-}$, will now be illustrated (*45, 46*). Direct reaction with glycine gives the bis-(*cis*)-glycinato Pt(II) complex

in 3:1 ratio with the *trans* isomer (*47*). If the above complex is treated with KOH in the presence of excess glycine then the complex Pt(glycine)₄ is obtained in which all the glycines are bound through their amino-groups only. It seems to be a general rule that the cyclic forms are obtained at neutral pH but are opened by both acid and alkali treatment (*48*).

Alanine, valine, and leucine, (amino-acids with alkyl substituents only) react in a manner very like that of glycine (*49—53*). All the reactions are rather slow and boiling solutions are normally employed in the preparative reactions. With the *cis*- and *trans*-isomers of Pt(NH₃)₂Cl₂ substitution of the chloride *only* occurs. Since the *trans*-labilising influence of the incoming groups of the amino-acids is very small, the −NH₂ groups remain stable. Consequently chelated complexes are only formed by the amino-acids in the case of the *cis*-isomer.

Histidine, an amino-acid with an imidazole group in its side-chain, reacts with $PtCl_4^{2-}$ slowly even in KOH and over some days (*54*). It gives the complex

By contrast methionine, −SCH₃ group in the side chain of the amino-acid, reacts rapidly with $PtCl_4^{2-}$ and both *cis*- and *trans*- Pt(NH₃)₂Cl₂ (*55, 56*). With the *trans*-isomer the following complex is formed which is stable in the presence of high concentrations of chloride ions and protons. The high thermodynamic stability of Pt—S bonds is apparent. With iodide ions however the methionine is rapidly lost (57). The ammonia ligands are not displaced in these reactions. However, in the reactions of the *cis*-isomer with methionine the high *trans*-labilising influence of the sulphur group becomes apparent in the following observed reaction scheme (58—60). Note that only the ammonia ligand *trans* to —S—CH$_3$ is labilised.

Parallel reactions are known with thio-urea (61).

A report (62) has appeared of the reaction of PtCl$_4^{2-}$ with one peptide, glycyl-glycine. No reaction takes place when the peptide is in the zwitter-ion form but in KOH the free aminogroup can displace chloride. No cyclic compounds were obtained presumable because of the large ring size demanded.

C. Interactions of Metal-complexes with Proteins

At pH 7 the following groups of proteins are likely to interact with metal ions,

$$-NH_2 \text{ (terminal or side-chain)},$$

$$-CO_2^- \text{ (terminal or side-chain)},$$

, $-CH_2 \cdot S^-$, -imidazole,

$$-CH_2-S-CH_3, \ -S-S-,\text{-phenolate}.$$

Protein-crystallography has revealed many of these groups as binding centers of heavy metals such as Pt(II), Hg(II), Au(III). It is important to note that the solution conditions in which the reagent has been applied to these crystals differ widely. If an acid pH has been used then groups with a high affinity for protons,

$$-NH_2, \ -C\diagup{}^{NH}_{NH_2} \ , \text{ and } -CH_2 \cdot S^-,$$

become unlikely candidates for attack. In chloride medium the concentration of chloride will prevent an incoming group from binding to the Pt in one of the planar positions. Finally [PtCl$_4$]$^{2-}$ seems best able to attack groups on the *surface* of proteins and it does not go to the more hydrophobic inner protein regions easily. Thus it is probably not such a general reagent as RHg.Cl. This has obvious implications as to the mode of action of the anti-tumour active Pt reagents: we suspect that they will attack the surface and not the inside of proteins. We now give some examples.

In the case of the protein, *lysozyme*, crystals were soaked in a chloride solution, pH 5.5 and [PtCl$_4$]$^{2-}$ 'floated' in. The anion bound between two arginines (14) of two different proteins. The procedure used, high chloride, ensured that PtCl$_4^{2-}$ remained intact and the binding is probably electrostatic. This is confirmed by the binding of PtBr$_4^{2-}$, PdI$_4^{2-}$ and HgI$_3^-$ to the same site. PdCl$_4^{2-}$ and PdBr$_4^{2-}$ bind in a different site close to histidine 15 and lysine 96. It could well be that these anions react by substitution with the histidine. The [PtCl$_4$]$^{2-}$ is not strongly bound in this protein. We are grateful to Dr. *C. C. F. Blake* (Oxford) for the above information.

In *ribonuclease-S* [PtCl$_4$]$^{2-}$ binds to the sulphur of methionine (63). In this case however greater interest centres on the use of cis-[PtenCl$_2$] which is one of the best anti-tumour agents. The conditions of application of the reagent were sulphate media at pH 5.5. This is a more favourable attacking medium than that used in the experiments with lysozyme. We quote results from the original paper (63).

"The closely related dichloroethylenediamine platinum(II) was found to be slightly better than PtCl$_4^{2-}$ as a heavy atom. A platinum-sulphur bond to methionine 29 is formed according to current analysis (now confirmed). The platinum cannot be removed by washing the crystal in sharp contrast to the reversible binding of other metal complexes."

The methionine 29 is on the *outside* of ribonuclease-S and simple absorbed [PtCl$_4$]$^{2-}$ or [PtCl$_2$en] should have been rapidly removed from this site. We therefore believe that the platinum complex has *reacted*

33

to give [enPt-methionine X] where the sulphur is bound in the plane and has displaced one chloride while some other protein group, may have displaced the other chloride. $[Pt(CN)_4]^{2-}$ which was used by the same authors does not bind at the same site suggesting that a nucleophilic displacement reaction has not taken place here in contrast with $[PtenCl_2]$. The strength of the attachment of $[PtenCl_2]$ was further demonstrated by a chromatographic run in which the Pt travelled with the protein, ribonuclease-S, through a whole separation procedure. These experiments establish methionine as a major centre of binding of $[PtCl_4]^{2-}$ and $[PtenCl_2]$ but there are many other sites of attack (see below). Neither lysozyme nor ribonuclease have free −SH groups. It would be quite wrong to conclude that any inhibitory process due to the addition of $[PtCl_2en]$ was due to methionine binding. The following further experiments on ribonuclease illustrate this point in part.

When the pH of ribonuclease was raised to 8.0 from 5.5 a second site became partially occupied in which the platinum bound to histidine 119. At the lower pH this group would be protonated and blocked. As stated above for this reason groups such as terminal −NH$_2$, arginines and lysines are not likely to be as effective as thio-ether in displacing chloride from chloride complexes of platinum. (We are grateful to Dr. *L. Johnson* (Oxford) for supplying us with much of this information).

In cytochrome-*c* $[PtCl_4]^{2-}$ binds to methionine 65 on the outside of the protein (*64*). Again the chloride is probably displaced as the reagent is applied in phosphate buffer. *Dickerson* and co-workers consider that the platinum is oxidized to a Pt(IV) complex but this seems unlikely as it does not occur in the reaction of $[PtCl_4]^{2-}$ with simple amino-acids.

In *carboxypeptidase A* a miscellany of derivatives have been prepared using $[PtCl_4]^{2-}$, Table 12 (*65*), and this is also true of chymotrypsin (*66*). As pointed out by *Dickerson et al.* (*64*) the major group to be attacked is methionine (in the absence of other sulphur groups) but the reagent is not as specific as they indicate. Table 12 shows now a variety of metals label carboxypeptidase and illustrate how one metal may label in many places. The reagent $[PtenX_2]$ has also been used by *Kraut* (*67*) and again it seems to label methionine particularly.

In total the analysis of protein complexes indicates that methionine is the most likely group to bind to $[PtCl_4]^{2-}$ and $[PtenCl_2]$ probably by a displacement reaction. However a *specific* attachment to another residue such as histidine or sulphydryl can not be ruled out. Furthermore the antitumour activity is that of *cis*-$[Pt(NH_3)_2Cl_2]$ and the di-functional character of this reagent has not been revealed so far by the above studies. In fact no attempt has been made to uncover the difference between $[PtCl_2(NH_3)_2]$ in its *cis* and *trans*-forms.

Table 12. *Specificity of heavy atom binding to carboxypeptidase (after W. N. Lipscombe)*

Heavy Atom	Residue (Amino-acid) Bound
Pb	Glutamate-270
Hg(1)	Histidine-69, Glutamate-72, Histidine-196
Hg(2)	Histidine-29
Hg(3)	Histidine-29, Lysine-84
Hg(4)	Histidine-303
Pt(1)	Cysteine-161
Pt(2)	Methionine-103
Pt(3)	Alanine-1 (N-terminus)
Pt(4)	Histidine-303
Ag(1)	Histidine-166, Serine-158
Ag(2)	Histidine-120
Ag(3)	Histidine-29, (Lysine-84)
Ag(4)	Histidine-303
Co(1)	Histidine-303
Co(2)	Histidine-29
Zn	Histidine-69, Glutamate-72, Histidine-196

The different metal derivatives are obtained under different initial solution conditions which leads to more than one derivative for a single metal.

We note in passing that *DNA-replicase* has an essential thiol group which is labelled by mercurials. To our knowledge no studies of the effect of platinum on the replicase have been made and this is clearly an omission which should be rectified. In some proteins platinum derivatives would seem to go for the same sites as mercury(II) reagents — a not surprising result as their chemistry is similar.

D. Interaction of Metal Ions with Nucleotides

The large number of metal ions which are known to bind to nucleic acids (*68*) can be divided into two broad categories depending upon their mode of interaction. The first group containing Na$^+$, K$^+$, Ca^{2+}, Mg^{2+}, Mn^{2+}, Co^{2+}, and Zn^{2+} bind very largely to the phosphate groups whereas a second class, largely the 'b'-metals and consisting of Hg^{2+}, CH$_3$Hg$^+$, Ag$^+$, and In^{3+} amongst others, Fig. 6, interact strongly with the purine and pyrimidine bases, Fig. 10. Cu^{2+} is apparently in an intermediate category binding to the phosphate at low metal to phosphorus ratios (*r*) but binding to the bases at higher ratios. The main evidence for the different modes of binding has been provided by potentiometric and spectrophotometric methods. On account of its similarity to the second group of metals platinum is expected to bind to DNA at the purine and pyrimidine bases and there is now some evidence that this is the case (*69, 70*).

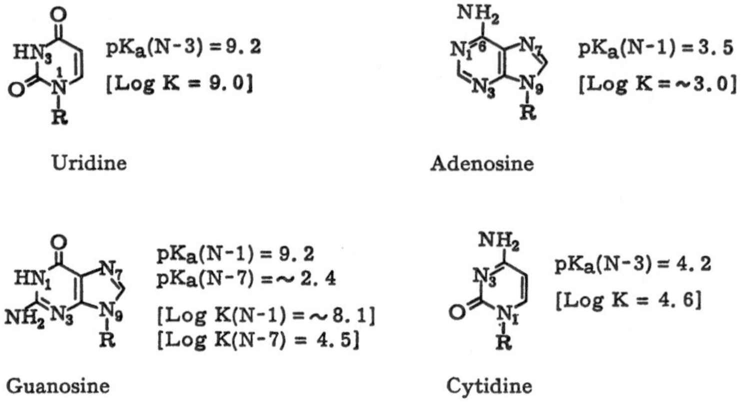

Uridine   Adenosine   Guanosine   Cytidine

$pK_a(N-3) = 9.2$

[Log K = 9.0]

$pK_a(N-1) = 3.5$

[Log K = ~3.0]

$pK_a(N-1) = 9.2$
$pK_a(N-7) = ~2.4$

[Log K(N-1) = ~8.1]
[Log K(N-7) = 4.5]

$pK_a(N-3) = 4.2$

[Log K = 4.6]

Fig. 10. The formulae of some of the bases in DNA and RNA. The acid dissociation constants and the methylmercury constants and binding centres are shown (after *R. B. Simpson*)

In this section we review briefly the manner in which the typical '*b*'-metals interact with DNA in particular and its components. We do not exclude the possibility that the important function of the platinum compounds is to react with RNA but at this stage of our knowledge this possibility introduces no new chemical complications. Most of the work to be reported with DNA and nucleotides and Hg^{2+}, Ag^+ and CH_3Hg^+ has been carried out by *Davidson (71—74)*, *Eichorn (75)* and *Simpson (76)* with their coworkers. On the other hand there is also a considerable amount of Russian work on the interaction of Ag^+, Au^{3+}, $PdCl_4^{2-}$, $PtCl_4^{2-}$ and the *cis*- and *trans*-isomers of both $Pd(NH_3)_2Cl_2$ and $Pt(NH_3)_2Cl_2$ with bases and especially the base analogues 5-fluorouracil and 6-mercaptopurine (*77—80*). Finally based upon the known chemistry of platinum we suggest some of the likely ways in which the anti-tumour agents may react with nucleic acids.

The binding sites of Hg^{2+}, Ag^+, CH_3Hg^+ are always the nitrogen atoms of the purine and pyrimidine bases. The preferences of the metals for these sites, in thermodynamic terms, parallel closely the *p*Ka values of the nitrogen centres, that is the thermodynamic binding of the proton to the same centres. The work of *Simpson (76)* on the binding of CH_3Hg^+ to the bases, see Fig. 10, reveals the parallel in a most striking manner.

The primary binding site of adenine is N(9) but when this is blocked as in adenosine it becomes N(1). This makes the binding of metals to adenosine weaker than the binding to N(7) of guanosine. The N(7) of guanosine is a known major site of alkylation too.

Binding of some '*b*'-metals also occurs at the 6-amino groups of adenosine and cytidine and with the 2-amino group of guanosine. Thus at higher pH values both Hg^{2+} and CH_3Hg^+ will bind at these sites displacing a proton (*76*). Silver would appear to require a still higher pH to attack the amino groups (*81, 82*). It is the high polarising power of the '*b*'-metals which brings about the ionisation of the amino-groups. Thus even with ammonia Hg^{2+} can form a compound with NH_2^-, $pK_a = 35$ (*83, 84*), while Ag^+ forms only a stable ammine. The pK_a of the 6-amino group of adenosine has been estimated to be 14 (*81*). Thus the displacement of a proton from the amino-group of a nucleotide is possible.

A further point of possible significance was revealed by Simpson's work. It was found that the binding constants to the amino-groups of both cytidine and adenosine were higher when N(3) or N(1) were blocked by metallation. Closely related to this effect is the change of pKa of the amino groups when these ring nitrogen positions are alkylated, protonated, or metallated. For example the pK_a of N(9) of 6-mercaptopurine decreases from 11.9 to about 8 on chelation of platinum to N(7) and the mercapto group. It is well-known that alkylation at N(7) of guanosine can lead to the scission of the glycosidic link at N(9) and hence in DNA to base deletions and chain rupture (*86*). The high polarising power of the '*b*'-metals could well cause similar damage to DNA for although this has not been observed we have stressed the similarity between the '*b*'-metals and carbonium ions in their general chemical effects.

When Hg^{2+}, CH_3Hg^+, and Ag^+ bind to DNA changes occur in the structure of the polymer which can be detected by measurement of such physical properties as viscosity, sedimentation and melting point curves but unfortunately these results are difficult to interpret. Spectral studies and the measurement of proton release on metal binding have been more informative. Ag^+ forms three types of complex (*73*). Type I is formed at values of r between 0 and 0.2 without release of protons. The binding is most important in DNA with a high CG content. Type II complexes appear at r values close to 0.5 and one proton is released per silver bound in the reaction. This complex is more favoured at high pH so that above pH about 8 type I and type II complex formation are not separable. It has been suggested by Davidson that type I binding is due to the formation of a π-sandwich complex between two aromatic rings of the same strand but it is not clear why the N(7) of guanosine is not implicated for this is the site for alkyl and CH_3Hg^+ binding. Type II binding is said to be between N(3) and N(1) of an AT pair or of a GC pair with the displacement of one hydrogen normally involved in hydrogen bonding. However the separations between these nitrogens is about 3.0 Å, Fig 11, while the N—Ag—N distance is normally 3.8 Å (85). Again the binding can be reversed readily by the addition of cyanide which is a little surprising if the DNA has been uncoiled by the action of Ag^+.

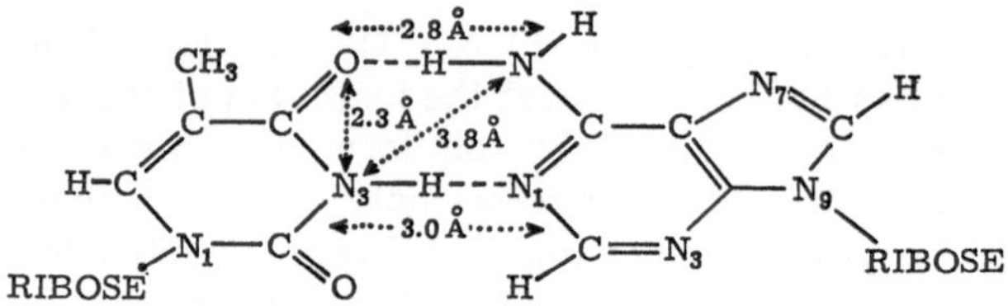

Fig. 11. The central structure of the DNA molecule

Hg^{2+} complexes readily with DNA. Only one type of complex is formed independent of the base ratio and up to an r value of 0.5. Two protons are released at pH 5—7 per metal bound. There is no satisfactory model for this binding. The removal of the Hg^{2+} by chloride reverses the changes in DNA in marked contrast with the irreversible changes induced by CH_3Hg^+. It may be that it is a special feature of a metal which can form two N—M—N bonds reversibly in DNA.

All the explanations of binding to DNA have ignored the possibility of binding to the amino-groups of the bases. These groups are readily accessible to metals and binding to them would not be expected to have much effect on the Watson-Crick structure. Protons are likely to be released from these groups as shown above so that it could well be that a good deal remains to be learnt about the binding of the metals to DNA.

The binding sites of platinum to DNA and its separate bases are expected to be similar to those for the other 'b'-metals, provided that the platinum binds through one site only. However both $PtCl_4^{2-}$ and such molecules as $Pt(NH_3)_2Cl_2$, cis- and trans-forms, can act bifunctionally. Indeed preferential binding is expected to yield chelated complexes of one type or another. For example the cis-chelation will lead to the following complexes.

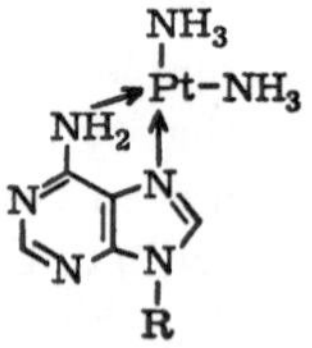

$$\text{-O} \rightarrow \text{Pt-NH}_3$$

The additional stability conferred by chelation, in spite of the poor proton accepting power of the 6-amino group of adenosine, may outweigh the stability of monodentate binding to N(1), the primary site for *trans*-[Pt(NH₃)₂Cl₂] and Hg⁺⁺. (The keto tautomer of guanosine are normally found in DNA and at *p*H 7, will have rather poor binding ability making adenosine the favoured base for bi-dentate chelation of this type).

Analogous compounds are well known in inorganic chemistry e.g.

Russian work confirms that *cis*- and *trans*-[Pd(NH₃)₂Cl₂] bind in the same manner to bases lacking a site of bi-dentate chelation. With 5-fluoro-uracil the *cis*-isomer yields

while the *trans*-isomer forms

However, when a bi-dentate site is available as in 6-mercaptopurine the following product is formed with the *cis*-isomer the ammonia ligands being displaced by the strongly *trans* labilizing —S— group of the incoming mercapto-purine. The following product forms from the *trans* isomer

With adenosine the *trans*-complex will undoubtedly react mono-functionally at the N(1) position, or, if that is blocked, at the N(7) position or the 6-amino group.

Thus a clear difference between the mode of binding of the *cis*- and *trans*-Pt(NH₃)₂Cl₂ emerges only when a bi-dentate site for chelation is available, as in the purines but particularly in adenosine. In the latter case this site involves the 6-amino group and the ring N(7). It is note-worthy that this site faces the center of the DNA helix but would be readily accessible to metal complexes. Furthermore it may be partic-ularly significant in the context of the present work that the N(7) site of guanosine is readily alkylated (by alkyl⁺) and alkyl⁺ is not unlike a cation such as [Pt(en)Cl]⁺ in its general electrophilicity. As stated above there are parallels between the chemistry of the '*b*' metals and carbonium cations. Since *trans*-[Pt(NH₃)₂Cl₂] can bind only monofunctionally at the N(7) site of purines, the stability of the resulting complex will be low (see above for Hg⁺⁺ at N(7)) compared with the chelate complex formed by *cis*-[Pt(NH₃)₂Cl₂].

Bidentate sites may be available in DNA itself. In the binding of Hg²⁺ and Ag⁺ we surmise that the binding is to amino-groups of cytidine and adenine and possibly the N(7) of guanine as all these positions are accessible in DNA. The introduction of a more kinetically permanent group such at [Pt(NH₃)₂]²⁺ attached to two such groups, e.g. two amino groups could lead to the formation of an inter-strand cross-link and an inhibition of DNA synthesis (*90*).

Another way in which Pt could bind to DNA is through the formation of intercalation compounds. The parallel here is with the hydrocarbon carcinogens and the nucleic acid stains, the acridines. It has been shown that metal chelates will form this same type of π-complex. For example, palladium oxinate will form exactly the same type of π-complexes as anthracene (*88*).

VI. Partition of Drugs

The *cis* and *trans* [Pt(NH$_3$)$_2$Cl$_2$] molecules are neutral. They are likely to pass through lipid membranes with comparative ease. The hydrolysis products [Pt(NH$_3$)$_2$(H$_2$O)$_2$]$^{2+}$ are unlikely to partition rapidly at low pH but as the pK$_a$ of the water molecules in the complex are ~5 and ~7, [Pt(NH$_3$)$_2$(OH$^-$)$_2$] is available and is neutral. Thus the action of the Pt compound, [Pt(NH$_3$)$_2$Cl$_2$] should be studied as a function of pH in order to discover the importance of partition between aqueous and non-aqueous solvents. Naturally the partition can be achieved by other changes in the NH$_3$ molecule. In practice it is possible to design *cis*-[Pt(NH$_2$R)$_2$Cl$_2$] complexes of any desired charge type. Now as drugs may act at the membrane or intracellularly the utilisation of differently charged complexes should permit an examination of the site of action.

VII. Conclusions — Mode of Action

It is not the purpose of this article to show how platinum compounds may act as drugs. We wish to indicate rather how knowledge of drug action may be more easily forthcoming using heavy metal atom compounds in place of organic molecules. Thus it could be that the causes of cancer (and possibly the cure) can be elucidated by the very detailed study of the action of *cis*-[Pt(NH$_3$)$_2$Cl$_2$] using a variety of probe techniques which are illustrated in the above sections. The tools for such an investigation are available or are becoming available and the advantages of using metals due to their probe properties could be very considerable. It might be thought that as heavy metals can bind to almost any group of reasonable donor strength they would be extremely unselective in a biological system. In fact this is not the case. Lead has a specific effect on porphyrin synthesis, cadmium accumulates in the kidney, mercury as methyl-mercury has an effect on the development of brain. The specificity can be seen at a molecular level by reference to the work of *Novick* (*89*) on gene mapping in bacterial DNA, or by an inspection of the highly specific ways in which metals are used in biopolymers.

Although bacteria have a single essential chromosome, one DNA molecule responsible for their genetic material, they may also contain extrachromosomal DNA. One variety of such DNA is viral DNA. This DNA can become incorporated into the chromosome and reproduced with it (lysogenic bacteria) or can become virulent — produced independently from the chromosome DNA — when it will cause lysis i. e. cell breakdown. The compound *cis*-[Pt(NH$_3$)$_2$Cl$_2$] causes the change from the lysogenic state to the virulent state in some unknown manner, Table 3.

Other forms of extrachromosomal DNA are known which can reproduce independently of the host DNA while they are not incorporated in it. Two major classes are the so-called R-factors and the plasmids. It should be clearly understood that these are physically independent self-replicating bits of DNA which cannot maintain the cell by themselves but often carry genes controlling resistance to various growth inhibitors. For example there are plasmids which carry genetic determinants for resistance to erythromycin and which carry penicillinase genes for destroying penicillin. In the last few years it has been found that the genes for resistance to many heavy metals are carried by the R factors and plasmids. Examples include resistance to arsenate, arsenite, antimony, cadmium, lead, mercury, indium, bismuth, zinc, and possibly cobalt and nickel.

The probable mode of action of the genes of the R-factors and plasmids is that they generate proteins which can protect against the metals presumably by strong metal binding. In any event the metal-inhibiting genes can be mapped on the plasmid DNA by conventional mapping methods much as is true for genes in chromosomal DNA. These genes are then found to map close to one another but there is a strong independence of protection against one metal, say lead, from all others. The only cases of metal markers coming close together (probably identical) are for Zn and Cd (see below) and for antimony and arsenite. The genetic map reads

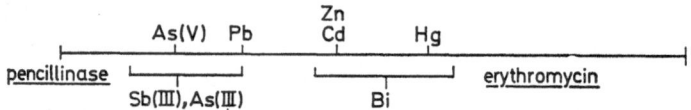

There does not seem to be any marker (other gene) in amongst those for the inorganic cations but it is not known how closely the genes are linked.

The picture of the plastid DNA as a special DNA carrying resistance genes to drugs is re-inforced in a striking though different way by these genes which specify different metal resistances. No corresponding genes for A-Group metals or for transition metals were found in these plastids although genes for resistance to cobalt and nickel have been described in *Esherichi Coli*. The quite remarkable grouping of the genes for protection against heavier B-Group metals may also occur in higher cells.

There is a protein, metallothionine, which is found in kidney and which binds cadmium and zinc very effectively. This may well be related to the bacterial protein. We see that *biological systems have developed highly selective ways of countering the influence of poisonous metals*. The protection involves the interaction between a selected protein and a given metal. We can now return to platinum chemistry.

The effect of platinum in a bacterial cell is to act in a very selective way — on cell division or causing lysis of lysogenic bacteria. It is likely that these changes are due to site specific attack on particular proteins or on particular bases in RNA or in DNA. It is necessary now to describe this attack in detail and to develop new probes for following the site *in vivo*. This exercise can be followed by a parallel examination of how *cis*-[Pt(NH₃)₂Cl₂] acts as an anti-tumour agent. Here we only point to some interesting observations.

1. The effects of the bifunctional Pt drugs are very like those of the cross-linking dialkylating agents known to be effective against cancer (*90*). The reader is reminded of the parallel chemistry of carbonium ions and of platinum complexes, section II F.

2. The effective compounds of platinum and the dialkylating agents both cause lysis in lysogenic bacteria, section I.

3. There is evidence that both classes of anti-tumour agent act by inhibiting DNA synthesis as opposed to RNA or protein synthesis. Many agents which inhibit DNA synthesis cause lysis of lysogenic bacteria-X-rays, ultra-violet light, many carcinogens (*3, 90*).

These facts alone would lead one to seek the origin of the effect of the platinum in its interaction with DNA. It should be possible to track this down with the aid of the physical methods of section III.

However it would be a great mistake to believe that all actions of metal drugs will be at the DNA level. The metal complexes described here act in a highly selective manner with proteins — this is why they are used to provide heavy atom derivatives for crystallographic work. Thus we may expect that there will be other effects of the heavy metals which are associated with RNA and protein interactions.

This paper originates with observations of Dr. *B. Rosenberg*. It was his stimulation that brought the authors together and they wish to record their indebtedness to him. All the views expressed here have been developed through discussion with him.

In September 1971 the seventh International Conference of Chemotherapy in Prague contained a full session on the use of platinum drugs. A set of papers will be published. Many of the points dealt with in passing in the above article are treated in detail confirming the general outline of the chemical factors which are of importance. The compound *cis*-Pt(NH₃)₂Cl₂ proved sufficiently successful in tests on tumours in mice that it is now being used in clinical trials in the United States of America.

A. J. Thomson, R. J. P. Williams, and S. Reslova

References

1. *Rosenberg, B., Van Camp, L., Trosko, J. E., Mansour, V. H.:* Nature *222*, 385 (1969).
2. — — *Krigas, T.:* Nature *205*, 698 (1965). — *Rosenberg, B., Van Camp, L., Grimley, E. B., Thomson, A. J.:* J. Biol. Chem. *242*, 1347 (1967).
3. *Reslova, S.:* to be published.
4. *Rosenberg, B., Renshaw, E., Van Camp, L., Hartwick, J., Drobnik, J.:* J. Bacteriol. *93*, 716 (1967).
5. *Bromfield, R. J., Dainty, R. H., Gillard, R. D., Heaton, B. T.:* Nature *223*, 735 (1969).
6. *Dwyer, F.:* In: Haemitin Enzymes,; eds. *Morton, R. A.,* and *Lemberg.* Oxford: Pergamon Press 1962.
7. For an account of transition metal chemistry see either *Cotton, F. A., Wilkinson, G.:* Advanced Inorganic Chemistry. New York: Wiley 1966. — *Phillips, C. S. G., Williams, R. J. P.:* Inorganic Chemistry, Vol. II. Oxford: Oxford Univ. Press 1966.
8. *Basolo, F., Pearson, R. G.:* Mechanisms of Inorganic Reactions. New York: Wiley 1958.
9. *Candlin, J. P., Taylor, K. A., Thompson, D. T.:* Reactions of Transition Metal Complexes. Amsterdam: Elsevier 1968.
10. *Basolo, F., Gray, H. B., Pearson, R. G.:* J. Am. Chem. Soc. *82*, 4200 (1960).
11. *Belluco, U., Cattalini, L., Basolo, F., Pearson, R. G., Turco, A.:* J. Am. Chem. Soc. 87, 241 (1965).
12. See references 8 and 9.: *Cattalini, L.:* In: Inorganic Reaction Mechanisms, p. 263; ed. *Edwards, J. O.* New York: Interscience 1971.
13. *Grinberg, A., Borzokova:* Russ. J. Inorg. Chem. (English Transl.) *2*, 2368 (1957) and *3*, 135 (1958).
14. *Williams, R. J. P.:* In: Transition Metal Chemistry, Vol. 2, p. 116; ed. *Carlin, R.* New York: Dekker 1966.
15. *Chatt, J., Gamlen, G. A., Orgel, L. E.:* J. Chem. Soc. 486, (1958).
16. *Day, P., Orchard, A. F., Thomson, A. J., Williams, R. J. P.:* J. Chem. Phys. *42*, 1973 (1965).
17. *Jorgensen, C. K.:* Absorption Spectra and Chemical Bonding in Complexes. Oxford: Pergamon 1962.
18. — Orbitals in Atoms and Molecules. London: Academic Press 1962.
19. *Chatt, J., Gamlen, G. A., Orgel, L. E.:* J. Chem. Soc. 1047 (1965).
20. *Linhard, Weigl:* Z. Anorg. Allgem. Chem. *226*, 49 (1951).
21. *Schatz, P. N., McCaffery, A. J.:* Quart. Rev. Chem. Soc. *33*, 552 (1969).
22. *Mason, S. F.:* Quart. Rev. Chem. Soc. *17*, 20 (1963).
23. *Blake, A., Peacocke, A. R.:* Biopolymers *6*, 1225 (1968).
24. *Hamilton, C. L., McConnell, H. M.:* Structural Chemistry and Molecular Biology; ed. *Rich, A., Davidson, N.* New York:: Freeman 1968.
25. *Gale, G. R., Howle, J. A., Smith, A. B.:* to be published.
26. *Shulman, R. G., Eisinger, J., Estrup, F. F.:* J. Chem. Phys. *43*, 3116, 3123, 3133 (1965).
27. *McDonald, C. C., Phillips, W. D.:* Biochem. Biophys. Res. Commun. *35*, 43 (1969).
28. *Morallee, K. G., Nieboer, E., Rossotti, F. J. C., Williams, R. J. P., Xavier, A. V., Dwek, R. A.:* Chem. Commun. 1132 (1970).
29. *Freeman, R., Murray, G. R., Richards, R. E.:* Proc. Roy. Soc. (London) *A* 242, 455 (1957).

30. *Packer, K. J.:* Progr. Nucl. Mag. Res. *3* (1967).
31. *Morallee, K. G., Williams, R. J. P., Woodworth, R.:* Biochemistry *9,* 839 (1970).
32. *Bryant, R. G.:* J. Am. Chem. Soc. *89,* 2496 (1967).
33. *Stengle, J. R., Baldeschwieler, J. D.:* Proc. Natl. Acad. Sci. U.S. *55,* 1020 (1966).
34. *Fancani, R., Tomlinson, B., Nafire, L. A., Small, W., Peticolas, W. L.:* J. Chem. Phys. *51,* 3993 (1969).
35. *Rimai, L., Kilponen, R. G., Gill, D.:* J. Am. Chem. Soc. *92,* 3824 (1970).
36. *Adams, D. M., Chatt, J., Gerratt, J., Westland, A. D.:* J. Chem. Soc. 734 (1964).
37. — *Chandler, P. J.:* J. Chem. Soc. 1009 (1967 A).
38. *Adamson, A. W., Waltz, W. L., Zinato, E., Watts, D. W., Fleischaner, P. D., Lindholm, R. D.:* Chem. Rev. *68,* 541 (1968).
39. *Perumareddi, J. R., Adamson, A. W.:* J. Phys. Chem. *72,* 414 (1968).
40. *Grinberg, A. A., Nikoleskaya, L. E., Bagesultanova, G. A.:* Dokl. Akad. Nauk. SSSR *101,* 1059 (1955).
41. *Babera, A. V., Mosyagina, M. A.:* Bull. Acad. Sci. USSR, Div. Chem. Sci. (English Transl.) 205 (1953).
42. *Balzani, V., Carassiti, V., Moggi, L., Scandola, F.:* Inorg. Chem. *4,* 1243 (1965). — *Scandola, F., Traverso, O., Balzani, V., Zuckini, G. L., Carassiti, V.:* Inorg. Chim. Acta *1,* 76 (1967).
43. — — *Scandola, F.:* Gazz. Chim. Ital. *96,* 1213 (1966).
44. *Renshaw, E., Thomson, A. J.:* J. Bacteriol. *94,* 1915 (1967).
45. *Grinberg, A. A., Inckova, E. N., Varswoky, C. Y.:* Russ. J. Inorg. Chem. (English Transl.) *8,* 1394 (1963).
46. *Volstejn, L. M., Zeghda, J. D.:* Russ. J. Inorg. Chem. (English Transl.) *13,* 833 (1968).
47. — — Russ. J. Inorg. Chem. (English Transl.) *7,* 5, 129, 1294, 1399 (1962).
48. — — Russ. J. Inorg. Chem. (English Transl.) *5,* 17 (1960).
49. — — Russ. J. Inorg. Chem. (English Transl.) *13,* 833 (1968).
50. — — Russ. J. Inorg. Chem. (English Transl.) *8,* 21 (1963).
51. — — Russ. J. Inorg. Chem. (English Transl.) *7,* 788, 1199 (1962).
52. — — Russ. J. Inorg. Chem. (English Transl.) *12,* 829 (1967).
53. — — *Anachova, L. S.:* Russ. J. Inorg. Chem. (English Transl.) *8,* 1072 (1963).
54. — *Lubyanova, I. G.:* Russ. J. Inorg. Chem. (English Transl.) *11,* 708 (1966).
55. — *Mogilevkina, M. F.:* Russ. J. Inorg. Chem. (English Transl.) *10,* 293 (1965).
56. — *Krylova, L. F., Mogilevkina, M. F.:* Russ. J. Inorg. Chem. (English Transl.) *12,* 832 (1967).
57. — — — Russ. J. Inorg. Chem. (English Transl.) *10,* 1077 (1965).
58. — *Mogilevkina, M. F.:* Russ. J. Inorg. Chem. (English Transl.) *8,* 304 (1963).
59. — *Krylova, L. F., Mogilevkina, M. F.:* Russ. J. Inorg. Chem. (English Transl.) *11,* 333 (1966).
60. — *Mogilevkina, M. F.:* Dokl. Akad. Nauk. (SSSR) *142,* 1305 (1962).
61. *Grinberg, A. A., Sevater, M., Gel'man, M. I.:* Russ. J. Inorg. Chem. (English Transl.) *13,* 1695 (1968).
62. *Volstejn, L. M., Motyagina, G. G.:* Russ. J. Inorg. Chem. (English Transl.) *10,* 721 (1965).
63. *Wyckoff, H. W., Hardman, K. D., Allewell, N. M., Inagami, T., Tsernoglori, D., Johnson, L. N., Richards, F. M.:* J. Biol. Chem. *242,* 3750 and 3984 (1967).
64. *Dickerson, R. E., Eisenberg, D., Varnum, J., Kopka, M. L.:* J. Mol. Biol. *45,* 77 (1969).
65. *Lipscomb, W. N., Hartsuck, J. A., Reeke, G. N., Quicho, F. A., Bethge, P. H., Ludwig, M. L., Steitz, T. A., Muirhead, H., Coppola, J. C.:* Brookhaven Symp. Biol. *21,* 24 (1968).

45

66. *Sigler, P. B., Blow, D. M., Matthews, B. W., Henderson, R.:* J. Mol. Biol. *35,* 143 (1968).
67. *Wright, C. S., Alden, R. A., Kraut, J.:* Nature *221,* 235 (1969).
68. *Eichorn, C.:* In: Bioinorganic Chemistry. Washington–New York: Am. Chem. Soc., Publ. No. 100, 1971.
69. *Mansy, S., Thomson, A. J.:* Unpublished observation.
70. *Drobnik, G., Horacek, P.:* Unpublished observation.
71. *Yamane, T., Davidson, N.:* J. Am. Chem. Soc. *83,* 2599 (1961).
72. — — Biochim. Biophys. Acta *55,* 609 (1962).
73. *Jensen, R. H., Davidson, N.:* Biopolymers *4,* 17 (1966).
74. *Gruenwedel, D. W., Davidson, N.:* J. Mol. Biol. *21,* 129 (1966).
75. *Eichorn, G. L., Clark, P.:* J. Am. Chem. Soc. *85,* 4020 (1963).
76. *Simpson, R. B.:* J. Am. Chem. Soc. *86,* 2059 (1964).
77. *Gel'fman, M. I., Kustova, N. A.:* Russ. J. Inorg. Chem. (English Transl.) *14,* 1113 (1969).
78. — — Russ. J. Inorg. Chem. (English Transl.) *14,* 110 (1969).
79. — — Russ. J. Inorg. Chem. (English Transl.) *15,* 47 (1970); *13,* 1230 (1968).
80. — — Russ. J. Inorg. Chem. (English Transl.) *14,* 985 (1969).
81. *Gillen, K., Jensen, R., Davidson, N.:* J. Am. Chem. Soc. *86,* 2792 (1964).
82. *Eichorn, G. L., Butzow, J. J., Clark, P., Tarion, E.:* Biopolymers *5,* 283 (1967).
83. *Bell, R. P.:* The Proton in Chemistry. London: Methuen 1959.
84. *Nijssen, L., Lipscomb, W. N.:* Acta Cryst. *5,* 604 (1952).
85. *Carey, R. B., Wyckoff, R. W. G.:* Z. Krist. *87A.* 264 (1934).
86. *Lawley, P. D.:* Progr. Nucl. Acid. Res. *5,* 89 (1967).
87. *Harder, H. E.:* Ph. D. Thesis. Michigan State University (1970).
88. *Williams, R. J. P., Wright, J. D., Prout, C. K.:* J. Chem. Soc. 1966 A, 747.
89. *Novick, R. P., Roth, C.:* J. Bacteriol. *95,* 1335 (1968). — *Novick, R. P.,* Bacteriol. Rev. *33,* 210 (1969).
90. *Loveless, A.:* Genetic and Allied Effects of Alkylating Agents. London: Butterworths 1966.

Received May 21, 1971

The Chemistry of Vitamin B₁₂-Enzymes*

J. M. Wood and D. G. Brown**

Department of Biochemistry, School of Chemical Sciences,
University of Illinois, Urbana, Illinois, USA

Table of Contents

* Some of the research reported in this review was supported by grants from the United States Public Health Service AM 12599, the National Science Foundation NSF GB 26593X, and the Dow Chemical Company.

** Present Address: Department of Chemistry, University of Idaho, Moscow, Idaho, USA.

47

I. Introduction

There is no other vitamin which exemplifies the interdependence between the human race and microorganisms better than Vitamin B_{12}. Mankind still relies for its very existence upon the capability of a few species of bacteria to synthesize this fascinating coordination complex. A concerted effort is being made in a number of laboratories to determine procedures for the chemical synthesis of this vitamin. However, even the most talented synthetic chemists would have to admit that the organic chemist's lament is applicable to Vitamin B_{12}:

"Oh Lord, I fall upon my knees
and pray that all my syntheses
may not be inferior
to those conducted by bacteria."

Apart from the research on the chemical synthesis of Vitamin B_{12}, some progress has been made on the biosynthesis of B_{12}-coenzymes; particularly in the laboratories of *Shemin* (1) and *Friedmann* (2).

In this review we shall not deal with the synthesis of this coordination complex, but we shall deal with the chemical properties of B_{12}-coenzymes with special emphasis on how these properties relate to B_{12}-enzyme mechanisms. Also, we shall show how B_{12}-catalyzed methyl-transfer reactions have special significance in the biosynthesis of methylated heavy metals in the aqueous environment, and how the synthesis of these organometallic compounds has special relevance to problems concerned with continuing global environmental health hazards.

The development of *magnetic resonance techniques* coupled with computer time averaging has made the study of enzyme structure and function by these techniques more fruitful. 'H NMR, ^{13}C NMR and ^{19}F NMR have been used successfully to determine the structure of B_{12}-compounds in solution. We are rapidly approaching the point where the structure and function of the B_{12}-coenzymes will be completely understood, and the need for the synthesis and study of simple B_{12}-model compounds such as the cobaloximes (3) will be no longer necessary. However, even though studies on the chemistry of B_{12}-coenzymes is a necessary prerequisite to our understanding of their biochemical role, it is a wrong assumption to expect that the chemical properties of free coenzymes in aqueous solution should be duplicated in the enzymes.

II. The History of B_{12}-Chemistry

In 1926 *Minot* and *Murphy* (4) demonstrated that pernicious anemia could be therapeutically controlled by the ingestion of whale liver.

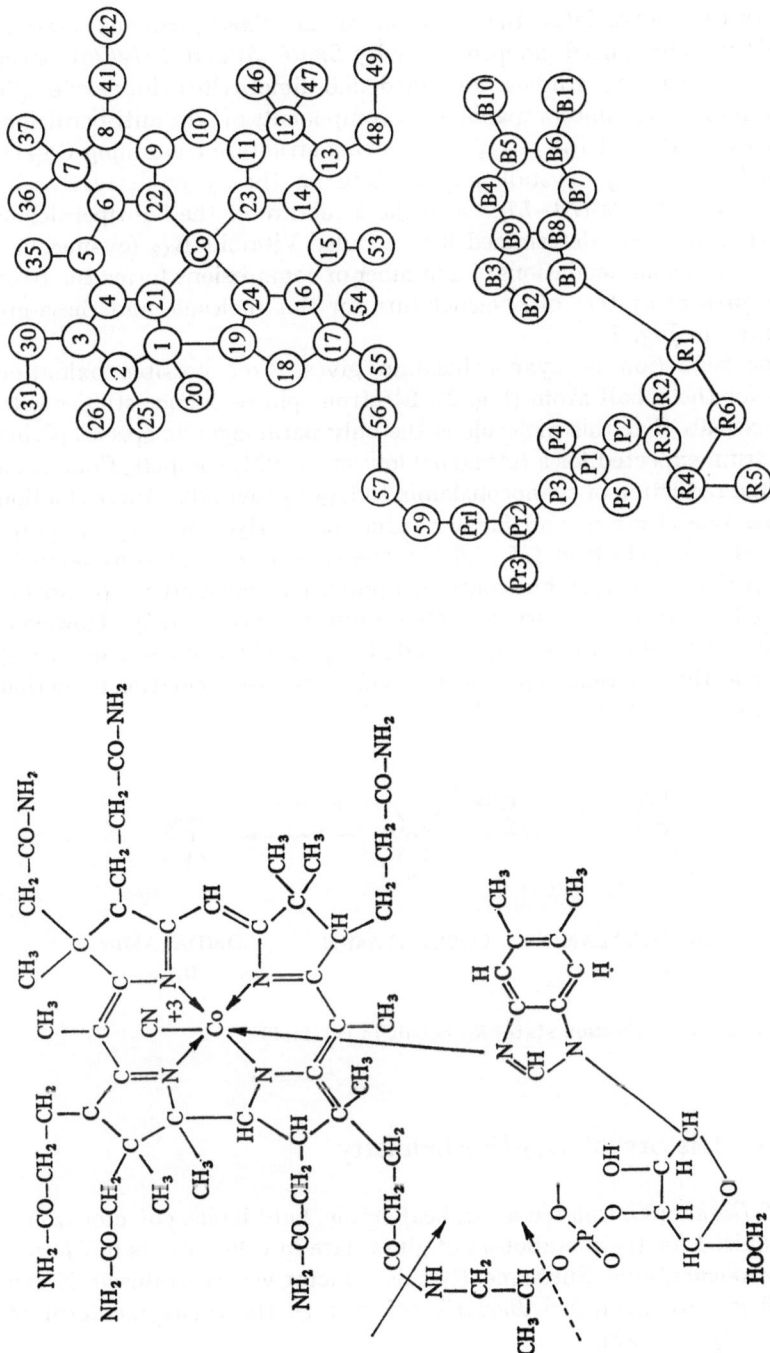

Fig. 1. Cyanocobalamin. (a) Hydrolysis at → gives cyanocobyric acid. (b) Hydrolysis at ----→ gives cyanocobinamide

Twenty-two years later the isolation of the *"anti-pernicious anemia factor"* was announced independently by *Smith* (5) and *Rickes* (6). Seven years of chemical studies identified 5,6-dimethylbenzimidazole (7), D-ribose (8) and amino-propanol (9) as components of "the anti-pernicious anemia factor", but the tetrapyrroline ring structure containing Co(III) awaited the X-ray crystallographic data on the cyano-derivative by *Hodgkin* and *White* (10—15). Once the structure of the "antipernicious anemia factor" was determined it was called Vitamin B_{12} (cyanocobalamin). The recommendations of a number of commissions forms the basis of the present system of nomenclature for this molecule and these are presented in Fig. 1.

The reduction of cyanocobalamin gives three possible oxidation states for the cobalt atom (Fig. 2). Electron spin resonance studies with B_{12}-r reveals that this molecule is the only paramagnetic species giving a spectrum expected for a tetragonal low spin Co(II) complex. Controlled potential reduction of cyanocobalamin to B_{12}-r proves that this reduction involves one electron, and further reduction of B_{12}-r to B_{12}-s requires a second single electron (16—19). At one time B_{12}-s was considered to be a hydride of Co(III), but controlled potential coulometry experiments provided evidence against a stable hydride species (16). However, these experimental data do not exclude the possibility of a stable Co(III) hydride as the functional species in enzyme catalyzed oxidation reduction reactions.

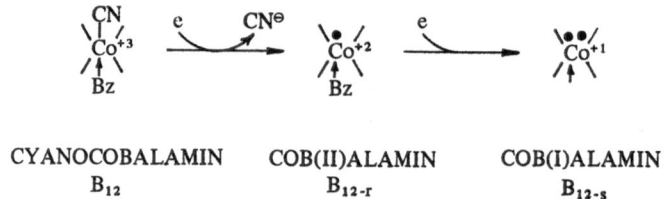

CYANOCOBALAMIN	COB(II)ALAMIN	COB(I)ALAMIN
B_{12}	B_{12}-r	B_{12}-s

Fig. 2. The three oxidation states for cobalt in corrinoids

III. The History of B_{12}-Biochemistry

In 1958 *Barker* (20) isolated a red, heat stable, light labile, cofactor which was required for the metabolism of glutamate in cell-extracts of *Clostridium tetanomorphum*. Subsequently this cofactor was crystallized. X-ray crystallography identified *Barker's* cofactor as the coenzyme form of Vitamin B_{12} (15, 21).

Lenhert and *Hodgkin* (*15*) revealed with X-ray diffraction techniques that 5'-deoxyadenosylcobalamin (B_{12}-coenzyme) contained a cobalt-carbon σ-bond (Fig. 3). The discovery of this stable Co—C-σ-bond interested coordination chemists, and the search for methods of synthesizing coenzyme-B_{12} together with analogous alkyl-cobalt corrinoids from Vitamin B_{12} was started. In short order the partial chemical synthesis of 5'-deoxyadenosylcobalamin was worked out in *Smith's* laboratory (*22*), and the chemical synthesis of methylcobalamin provided a second B_{12}-coenzyme which was found to be active in methyl-transfer enzymes (*23*). A general reaction for the synthesis of alkylcorrinoids is shown in Fig. 4.

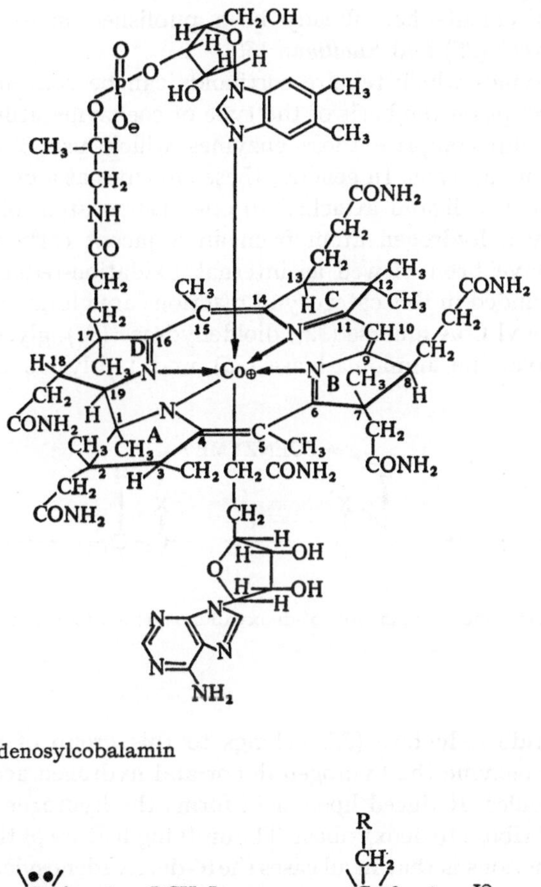

Fig. 3. 5'-deoxadenosylcobalamin

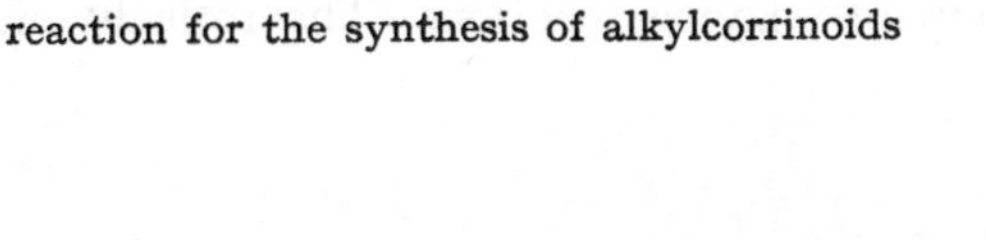

Fig. 4. General reaction for the synthesis of alkylcorrinoids

It soon became apparent that the biologically active forms of Vitamin B_{12} contained the unique Co–C-σ-bond, and the instability of these covalent compounds to visible light facilitated observations on the occurrence of functional corrinoids in a number of enzymes. Deoxyadenosylcobalamin was found to be the most abundant corrinoid in bacteria (24) and in mammalian liver (25). Methylcobalamin was found in *Escherichia coli* (26), calf liver and human blood plasma (27), and also in a number of Clostridia (28).

Following the elucidation of the structure of the biologically active forms of Vitamin B_{12} in 1961, a number of enzymes have been well characterized which require B_{12}-coenzymes. A survey of the properties of these enzymes in terms of molecular weight, subunits and other cofactor requirements has already been published in excellent reviews by *Hogenkamp* (29) and *Stadtman* (30).

The enzymes which require corrinoids can be conveniently divided into two groups on the basis of the type of coenzyme utilized. The most common group comprise those enzymes which use 5′-deoxyadenosyl corrinoids for catalysis. In general, these enzymes as a group involve the replacement of a ligand attached to one carbon atom of the substrate molecule by a hydrogen atom from an adjacent carbon atom. These reactions have been viewed as internal oxidation-reduction reactions (Fig. 5). Included in this category of reactions are glutamate mutase (31) methylmalonyl CoA mutase (32), dioldehydrase (33), glycerol dehydrase (34), ethanolamine ammonia-lyase (35) und L-β-lysine isomerase (36).

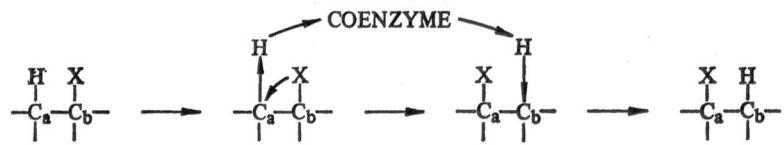

Fig. 5. General mechanism for 5′-deoxyadenosylcorrinoid catalyzed hydrogen transfer

Ribonucleotide reductase (37) belongs to this group of reactions also, but for this enzyme the hydrogen donor and hydrogen acceptor are different molecules. Reduced lipoic acid forms the hydrogen donor for the reduction of ribose to deoxyribose. The unifying feature of these seemingly different reactions is that in all cases the 5′-deoxyadenosylcorrinoid serves as intermediate carrier for an intramolecular hydrogen transfer.

Methylcobalamin acts as the functional molecule for methyl-transfer in a second group of enzyme reactions. Theoretically methyl-transfer from this coenzyme may proceed by three alternative routes (Fig. 6). Methylcorrinoids function as methyl-donor for the enzymatic synthesis of methionine (38), methane (39), and for the methyl group of acetic acid (40).

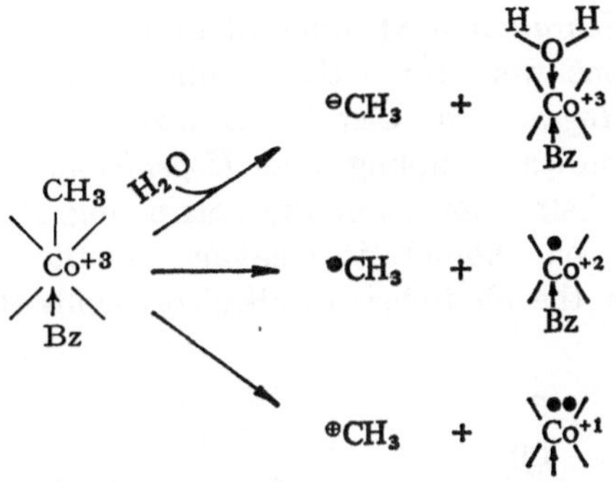

Fig. 6. Alternative mechanisms for the transfer of methyl groups from methylcorrinoids

IV. Properties of Alkylcorrinoids of Biological Significance

A documentation of the physical and chemical properties of alkylcorrinoids and alkyl-cobalt B_{12} model compounds is presented in a recent review by *Hill* (41). In this report we shall deal with only those properties of alkylcorrinoids which have proved to be useful when applied to a study of B_{12}-enzyme mechanisms.

The photochemistry of alkylcorrinoids has been used to facilitate our understanding of the involvement of the Co—C-σ-bond in enzyme reactions. Photolysis of 5'-deoxyadenosylcobalamin in the presence of oxygen yields aquocobalamin and adenosine-5'-aldehyde as the major products. However, when photolysis is conducted under anaerobic conditions, the products are Cob(II)alamin (B_{12r}) and 8, 5'-cycloadenosine (42). These data indicate that homolytic cleavage of the Co—C-σ-bond occurs, and further confirmation for homolysis was obtained by studying the photolysis of methylcobalamin (43). *Hogenkamp* (44) studied the photolysis of ^{14}C-methylcobalamin in aqueous solution with argon as the gas phase and found that methane plus ethane were the major products. The specific activity of the ^{14}C-methane was identical to that of the ^{14}C-methylcobalamin used for this photolysis, but the specific activity of the ethane was half that of the methane. These data indicate that homolysis of the Co—C-bond occurs and methane plus ethane are formed by hydrogen or methyl abstraction by methyl-radicals. Two years later *Schrauzer* (45) reinvestigated the photolysis of methylcobalamin, and reported markedly different results from those of *Hogenkamp*. Using CD_3-methylcobalamin and mass spectrometry *Schrauzer* concluded that methane was formed by a reductive mechanism and that ethane was produced by radical coupling because his products were $CD_3H + C_2D_6$. It should be pointed out that *Hogenkamp* and *Schrauzer* used different light, and *Schrauzer's* methylcobalamin solutions were three times more concentrated than those photolysed by *Hogenkamp*. Clearly, *Schrauzer's* reaction conditions would promote radical coupling and *Hogenkamp's* conditions would promote methyl abstraction from the corrin ring. Despite this discrepancy it is clear that homolytic cleavage of the Co—C-bond is the initial reaction for the photolysis of alkylcorrinoids (Fig. 7).

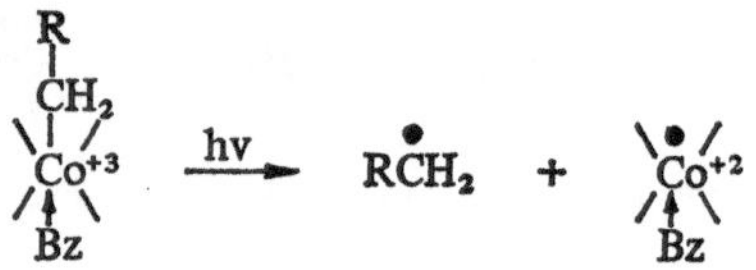

Fig. 7. General mechanism for photolysis of alkylcorrinoids

Although methylcobalamin is not susceptible to nucleophilic displacement by $CN^\ominus$-, 5'-deoxyadenosylcobalamin is converted to adenine and the cyanohydrin of 2,3-dihydroxypentenal. Therefore, it is possible that nucleophilic displacement of the 5'-deoxyadenosyl ligand could occur at the catalytic site of enzymes which require 5'-deoxyadenosylcorrinoids. Electrophilic attack by mild acid hydrolysis, iodine, or mercuric ions has been reported for alkylcobalamins. Mercuric ion has been shown to be a better electrophile for methyl groups on cobalt than either H^+ or I_2 (*46—49*). The reaction of mercuric ion with methylcobalamin to give aquocobalamin and methylmercury can be easily followed by 220 MHz NMR (Fig. 8). Kinetic studies performed with this reaction have shown that the rate of the reaction is a function of pH and Hg^{2+} concentration. Demethylation of methylcobalamin is the rate limiting step and is a second order rate process. However, as the reaction approaches stoichiometry, pseudo first order kinetics are observed (*49, 50*). The change in the kinetics of this reaction is due to CH_3Hg^+ functioning as an electrophill for methylcobalamin giving dimethylmercury as a product (Fig. 9). The synthesis of methylmercury compounds from methylcorrinoids and mercuric ion provides one example of how microorganisms can convert inorganic heavy metals into deadly poisonous organometallic compounds (*46*). This reaction is of great significance in problems of environmental pollution by heavy metals.

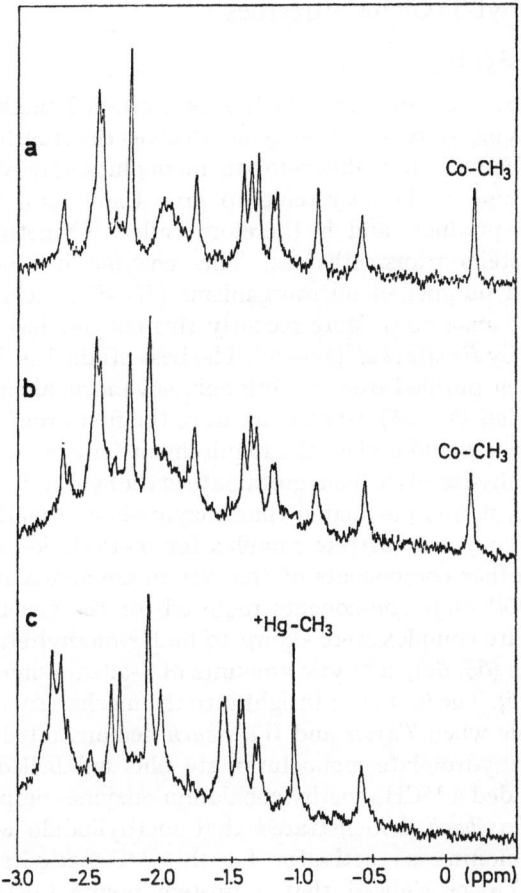

Fig. 8. 220 MH*z* MMR spectra of (a) methylcobalamin (b) methylcobalamin $+$ Hg_2^{2+} (c) methylcobalamin $+$ Hg^{2+}

$$H_2O \;+\; \overset{CH_3}{\underset{Bz}{Co^{+3}}} \;+\; Hg^{+2} \;\rightleftharpoons\; \overset{CH_3}{\underset{\underset{BzHg^+}{H\;O\;H}}{Co^{+3}}} \;\longrightarrow\; \overset{H\;O\;H}{\underset{Bz}{Co^{+3}}} \;+\; CH_3Hg^+$$

Fig. 9. Mechanism for the synthesis of methylmercury from Hg^{2+} and methyl-cobalamin

V. The Methyl-Transfer Enzymes

1. Methionine Synthetase

The biosynthesis of methionine is the best studied methylcobalamin-dependent enzyme system. This enzyme involves the transfer of a methyl group from N^5-methyltetrahydrofolate monoglutamate via a methyl-cobalamin-enzyme to homocysteine to give the essential amino acid methionine as product, and is therefore called N^5-methyltetrahydro-folate-homocysteine-transmethylase. This enzyme has been partially purified from a number of microorganisms (51—53), including aerobes, and facultative anaerobes. More recently this enzyme has been purified from hog liver by *Brodie et al.* (54—56). The best studied methionine synthetase has been purified from a methionine-cyanocobalamin auxotroph of *Escherichia coli* (57—58). *Guest et al.* were the first group to show that methylcobalamin would replace the requirement for cyanocobalamin and N^5-methyltetrahydrofolate monoglutamate in the synthesis of methionine (59). Therefore, it was postulated that enzyme-bound methylcobalamin represents the enzyme-substrate complex for methylation of homocyst-eine, and the other components of the system are necessary to generate this complex (60—62). Components required for the generation of this enzyme-substrate complex were shown to be N^5-methyltetrahydrofolate monoglutamate (63, 64), catalytic amounts of S-adenosylmethionine (65) and $FADH_2$ (65). The first clear insight into the mechanism of methionine synthetase came when *Taylor* and *Weissbach* demonstrated that $^{14}CH_3$-N^5-methyltetrahydrofolate monoglutamate plus unlabelled S-adenosyl-methionine yielded a $^{14}CH_3$-methylcobalamin-enzyme complex (66). Also *Taylor* and *Weissbach* demonstrated that methyliodide would replace S-adenosylmethionine as activator for this ^{14}C-methyltransfer. Both *Jaenicke* and *Taylor* showed that a protein bound Co(I)-species was the acceptor of CH_3^+ from N^5-methyltetrahydrofolate in this reaction (67, 68) (Fig. 10). Subsequent transfer of the methyl group as CH_3^+ from enzyme bound methylcobalamin to homocysteine regenerates the Co(I)-enzyme for remethylation from N^5-methyltetrahydrofolate. The role of S-adenosylmethionine or methyl iodide is simply to activate the enzyme by ensuring that it is fully methylated. It is believed that under normal physiological conditions such activation would not be necessary. *Taylor* and *Weissbach* believe that the inactive form of methionine synthetase has sulfur as an axial ligand because hot ethanol extraction of the cobal-amin from inactive enzyme yields a compound with spectral properties identical to sulfitocobalamin (69).

To summarize these data it appears that this enzyme system involves CH_3^+-transfer to a Co(I)-enzyme to give a methylcorrinoid-enzyme complex. Subsequent transfer of CH_3^+ from this methylcorrinoid-enzyme

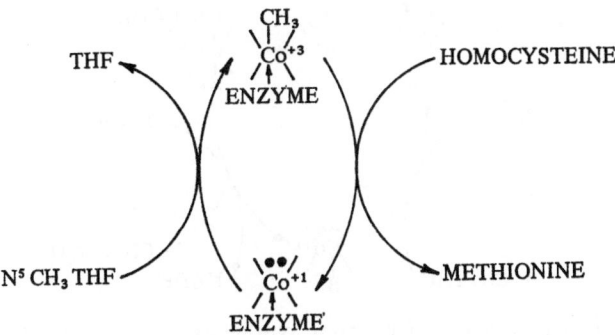

Fig. 10. Proposed mechanism for methionine synthetase

complex to homocysteine would regenerate a Co(I)-enzyme. One peculiar result remains to be answered in this enzyme system, and that is the observation that the CH_3-Co-σ-bond of the methylcorrinoid-enzyme complex is stable to light. We believe that this phenomenon is due to isomerization of the corrin ring by the methionine synthetase apoenzyme. Further support for this contention will be presented in the section on NMR.

2. Acetate Synthetase

Clostridium thermoaceticum is capable of incorporating $^{14}CO_2$ into both carbon atoms of acetic acid (*70, 71*). Direct proof for this unusual reaction came from mass spectrometry studies by Wood using $^{13}CO_2$ (*70*). When $^{14}CH_3$-methylcobalamin was shown to be the precursor of C_2 of acetic acid it became possible to consider a mechanism for this reaction (*72, 73*). Subsequently *Ljungdahl et al.* (*74*) isolated a carboxymethylcorrinoid from whole cells of *Clostridium thermoaceticum*, and when it was determined that acetic acid could be synthesized from carboxymethylcobalamin by an enzyme system requiring NADPH, then a mechanism could be proposed for this synthesis (Fig. 11) (*75*).

In this reaction sequence acetic acid synthesis requires methyl transfer as CH_3^+ to a Co(I)-corrin by N^5-methyltetrahydrofolate monoglutamate to give a methylcorrinoid intermediate which is carboxylated to give a carboxymethylcorrinoid. This carboxymethylcorrinoid would then be reductively removed by NADPH to give acetic acid and regenerate the Co(I)-corrin. In contrast to the methyl-transfer proposed for the methionine synthetase reaction, this mechanism suggests that CH_3^- stabilized by Co attacks CO_2 to give a carboxymethylcorrinoid intermediate.

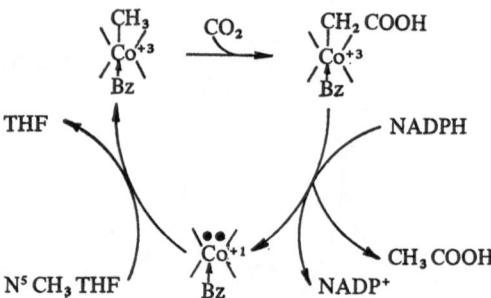

Fig. 11. Proposed mechanism for acetate synthetase

3. Methane Synthetase

Methanogenic bacteria contain considerably higher concentrations of corrinoids than other anaerobic bacteria (76). First *Blaylock* and *Stadtman* (77), then *Wolin et al.* (78) demonstrated that methylcobalamin is an excellent substrate for methane formation by cell-extracts of *Methanosarcina barkerii* and *Methanobacillus omelianskii* respectively. *Lezius* and *Barker* (76) isolated corrinoids from *Methanobacillus omelianskii*, and although very little methylcorrinoids were isolated from these bacteria, Factor III corrinoids were characterized. (Factor III is cobalamin in which 5,6-dimethylbenzimidazole is replaced by the base 5-hydroxybenzimidazole). Further evidence for the involvement of methylcorrinoids in methane synthetase was obtained when *Wood* and *Wolfe* (79), and *Blaylock* (80) implicated methyl-Factor III as a component of methane synthetase in *Methanobacillus omelianskii* and *Methanosarcina barkerii*. It is of interest to note that methyl-Factor III was first isolated from methanogenic sewage sludge by *Friederich* and *Bernhauer* (81). A reappraisal of the work with *Methanobacillus omelianskii* was found to be necessary when it was discovered that the original culture isolated by *Barker* (82) consists of a symbiotic culture of two organisms (83).

a) S-organism fermentation

$$CH_3-CH_2-OH + H_2O \longrightarrow CH_3-COOH + 2\,H_2$$

b) M.O.H.-organism fermentation

$$4\,H_2 + CO_2 \longrightarrow CH_4 + 2\,H_2O$$

This symbiotic relationship between H_2-evolver (S-organism) and H_2-utilizer (M.O.H.-organism) provides close interdependence in which M.O.H.-functions in methane synthesis. Cell extracts of both the symbiotic culture *Methanobacillus omelianskii* and methanogenic organism M.O.H. form methane from CO_2 (*84*), C_3 of serine (*85*), C_1 of pyruvate (*84*), N^5-methyltetrahydrofolate monoglutamate (*86*) and a variety of methylcorrinoids including methylcobalamin (*78*) methylcobinamide (*87*) and methyl-Factor III (*87*). Recent experiments with $^{14}CH_3$-methylcorrinoids indicate that the methyl group is not reductively displaced from cobalt in this enzyme system, but is transferred to a second coenzyme (M) of unknown structure before reductive cleavage of CH_3-(M) to methane (*88*). Furthermore, kinetic studies with CD_3-cobalamin and CT_3-cobalamin (*89*) supported by evidence for fluoromethyltransfer from CF_2H-cobalamin in methane synthetase (*90*) strongly suggest that methyl-transfer to coenzyme (M) occurs by transfer of a methyl radical ($CH_3^{\cdot}$). Until this enzyme system is purified it is not possible to propose a mechanism for methane formation. Many questions remain unanswered especially the role of catalytic amounts of ATP (*91*), and the source of electrons for reduction of methyl groups from coenzyme (M) to give CH_4. However, the synthesis of methylcorrinoid from N^5-methyltetrahydrofolate and the transfer of $CH_3^{\cdot}$ to coenzyme (M) has good experimental support (Fig. 12).

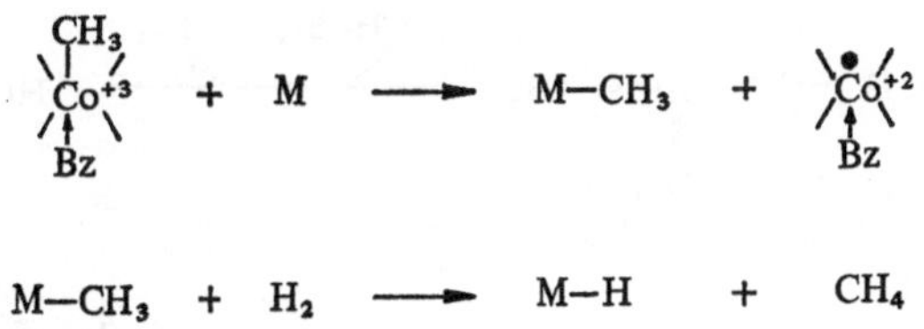

Fig. 12. Proposed mechanism for methane synthetase

4. Methyl-Transfer to Mercury

A survey of the methylating agents in biological systems reveals that there are three major compounds involved in this reaction:

1. S-adenosylmethionine
2. N^5-methyltetrahydrofolate derivatives
3. Methylcorrinoid derivatives

S-adenosylmethionine and N^5-methyltetrahydrofolate derivatives are not capable of transferring methyl groups to mercury salts since for both these coenzymes the methyl group is transferred as CH_3^+.

After *Halpern* and *Maher* (*92*) demonstrated that methylpentacyanocobaltate would react with mercuric ions to give methylmercury as the product, *Wood et al.* (*46*) demonstrated that this reaction occurred with methylcobalamin both chemically and enzymatically to give methyl-mercury and dimethylmercury as products. The kinetics for this reaction have now been worked out in detail, and rate constants for the formation of a mercuric-methylcobalamin intermediate and the conversion of this intermediate to aquocobalamin and methylmercury determined (*89*).

5. Methyl-Transfer to Arsenic

In 1933 *Challenger et al.* discovered that trimethylarsine was synthesized from inorganic arsenic compounds by molds (*93*). Recently, *McBride* and *Wolfe* (*94*), have reported the synthesis of dimethylarsine from arsenate by cell extracts of the methanogenic bacterium M.O.H. Methylcobalamin is the alkylating coenzyme for this synthesis which requires reduction of arsenate to arsenite, methylation of arsenite to methylarsonic acid, reduction and methylation of methylarsonic acid to dimethylarsinic acid, and finally a four electron reduction of dimethylarsinic acid to dimethylarsine (Fig. 13).

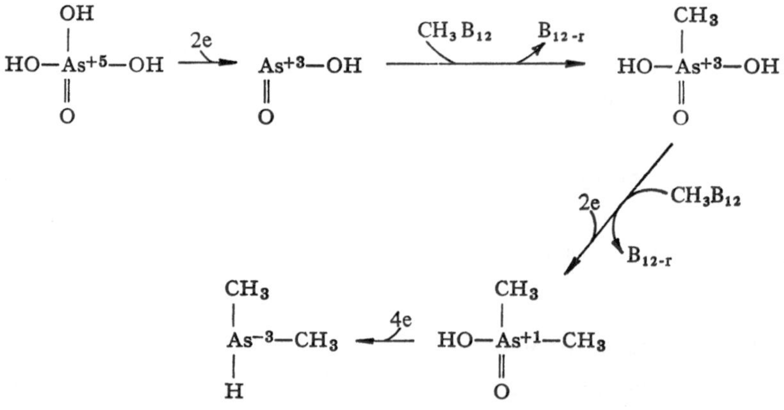

Fig. 13. Mechanism for the synthesis of dimethylarsine from arsenate and methyl-cobalamin

Clearly pollution hazards exist when inorganic arsenic compounds are introduced into an environment where anaerobic bacteria are growing. Arsenic impurities in the phosphate used in detergents and for agricultural practices may pose serious problems because of the continuing synthesis of deadly poisonous methylarsenic compounds.

VI. The Hydrogen-Transfer Enzymes

1. Hydrogen-Transfer and C—C Bond Cleavage

Although there is a great deal of similarity between the mechanism of intra-molecular hydrogen transfer via 5′-deoxyadenosyl corrinoids in the enzymes there is some variety for the nature of the bond which is labilized during these internal oxidation reduction reactions. For glutamate mutase, methylmalonyl CoA mutase and α-methylene glutamate mutase, cleavage of a C—C-bond is necessary in order to facilitate isomerization (Fig. 14). Both hydrogens from the 5′-CH_2-group of the 5′-deoxyadenosyl ligand to the Cobalt atom can be transferred in these reactions (30). Little exchange between free and bound 5′-deoxyadenosylcobalamin occurs during catalysis which indicates that a single molecule of co-enzyme will participate in thousands of catalytic reactions. With the exception of ribonucleotide reductase little or no exchange between free and bound 5′-deoxyadenosylcobalamin occurs for these corrin enzymes. Although the enzymes which involve C—C-bond cleavage must be very similar mechanistically there is a difference between methylmalonyl CoA

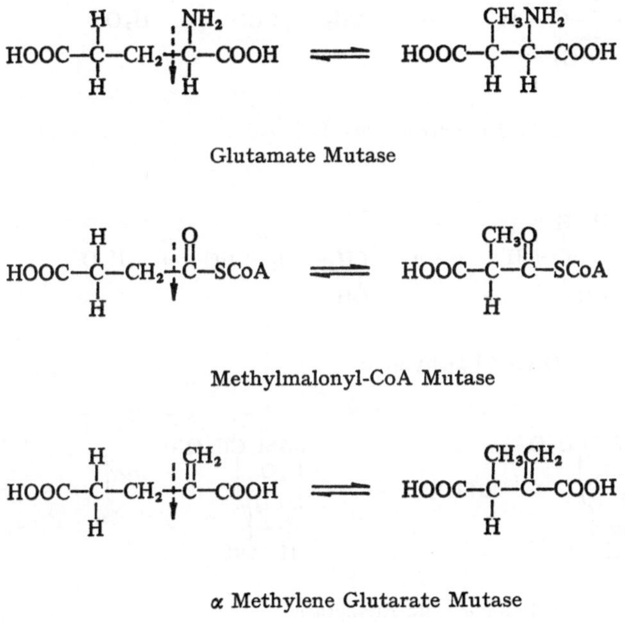

Glutamate Mutase

Methylmalonyl-CoA Mutase

α Methylene Glutarate Mutase

Fig. 14. Hydrogen-transfer with C—C bond cleavage

mutase and glutamate mutase in one respect. The isomerization of methylmalonyl CoA to succinyl CoA involves retention of configuration at C_2. For the conversion of L-threo-β-methylaspartate to L-glutamate inversion of configuration at C_3 of β-methylaspartate occurs. Therefore, the hydrogen migrating from the methyl group of methylaspartate inserts at C_3 at the opposite side to the departing glycyl group. Details of the molecular weights, and other cofactor requirements for these enzymes are documented in detail by *Stadtman* (*30*) and *Hill* (*41*).

2. Hydrogen Transfer and C—O Bond Cleavage

The enzymes which fit into this group include diol dehydrase, glycerol dehydrase and ribonucleotide reductase (Fig. 15). The conversion of propane 1,2 diol to propionaldehyde by diol dehydrase has been studied in detail by *Abeles et al.* (*95—98*). There is no exchange of hydrogen with solvent, and no enzyme free intermediates are involved. Hydrogen is transferred intramolecularly via the 5'-CH_2-group of 5'-deoxyadenosyl-cobalamin. Hydrogen migration from C_1 to C_2 of propane 1,2-diol proceeds with inversion of configuration at C_2. Studies with ^{18}O-labelled

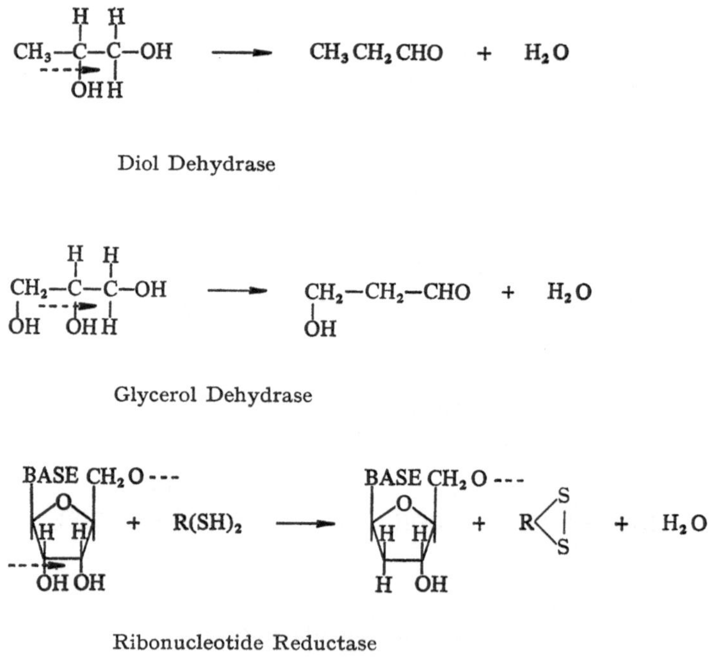

Diol Dehydrase

Glycerol Dehydrase

Ribonucleotide Reductase

Fig. 15. Hydrogen-transfer with C—O bond cleavage

propane 1,2-diol indicated that hydrogen transfer from C_1 to C_2 is accompanied by OH-transfer from C_2 to C_1. Recent experiments with diol dehydrase indicate that a Co(II)-corrinoid is an intermediate during catalysis, and this observation supports a free radical mechanism for hydrogen transfer. Glycerol dehydrase functions by a similar mechanism to diol dehydrase.

Ribonucleotide reductase differs from the other 5'-deoxyadenosyl-cobalamin requiring enzymes in a number of respects. Hydrogen is transferred from coenzyme to the C_2'-position of the ribose moiety without inversion of configuration. Also since lipoic acid functions in hydrogen transfer, exchange with solvent protons takes place. Furthermore, exchange between free and bound 5'-deoxyadenosylcobalamin occurs rapidly during catalysis. Evidence for a Co(I)-corrin as an intermediate for this reduction is presented in our section on electron spin resonance.

3. Hydrogen-Transfer and C—N Bond Cleavage

Ethanolamine ammonia lyase, L-β-lysine mutase, D-α-lysine mutase and ornithine mutase are representative of cobamide enzymes in which transfer of hydrogen occurs with cleavage of the C—N bond (Fig. 16).

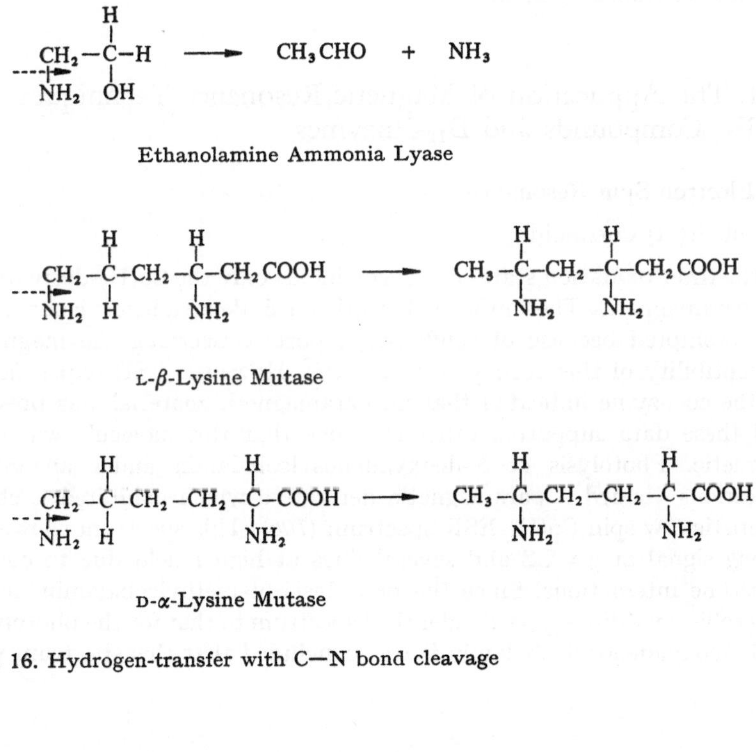

Ethanolamine Ammonia Lyase

L-β-Lysine Mutase

D-α-Lysine Mutase

Fig. 16. Hydrogen-transfer with C—N bond cleavage

Ethanolamine ammonia lyase has a molecular weight of 520,000 and consists of 8 or 10 subunits. Two 5′-deoxyadenosylcobalamin molecular bind per enzyme molecule, and recent kinetic studies by *Babior* show that these two molecules carry out catalysis independently. Evidence is available that this enzyme functions by a radical mechanism since both spin labeling and Co(II) esr experiments indicate that Co(II) is an intermediate during H-transfer. Also, 5′-deoxyadenosine has been detected as a product of oxygenation of the enzyme-substrate complex (*99—101*).

A most interesting combination of coenzymes is involved in the D-α-amino lysine mutase reaction. Pyridoxal phosphate is required for activity, and this coenzyme is probably involved in the transfer of the NH_2-group from C_6 to C_5 of the lysine molecule. Also, as with ribonucleotide reductase, the hydrogen at C_6 exchanges with solvent. L-β-lysine mutase does not require pyridoxal phosphate, but does require pyruvate, and this keto-acid is probably involved in -NH_2-transfer. Both of the lysine mutases require a sulfhydryl protein for catalytic activity. *Stadtman* (*30*) has pointed out that many cobamide enzymes have a requirement either for a sulfhydryl protein or a sulfhydryl containing coenzyme such as reduced lipoic acid. The interaction of the thiol group in these enzyme systems has not been investigated thoroughly. However, it would be unusual if this requirement does not play an important role in these H-transfer enzymes.

VII. The Application of Magnetic Resonance Techniques to B_{12}-Compounds and B_{12}-Enzymes

A. Electron Spin Resonance

1. Cobalt(II) Corrinoids

Of the three oxidation states of the corrinoids only the Co(II)-derivatives are paramagnetic. The initial ESR work on 5′-deoxyadenosylcobalamin was prompted because of conflicting reports concerning the magnetic susceptibility of this coenzyme (*102—105*). However, ESR experiments on the coenzyme indicated that no paramagnetic material was present and these data supported other evidence that this molecule was diamagnetic. Photolysis of 5′-deoxyadenosylcobalamin under anaerobic conditions yielded a paramagnetic derivative which exhibited a characteristic low spin Co(II) ESR spectrum (*106*). This spectrum showed a strong signal at $g = 2.2$ and several lines at higher field due to cobalt hyperfine interactions. Since the photolysis of methylcobalamin under anaerobic conditions gave an identical spectrum to that for the photolysis of 5′-deoxyadenosylcobalamin it was concluded that this spectrum was representative of Cob(II)alamin (B_{12-r}) with 5,6-dimethylbenzimidazole coordinated (*107*). Further support for this conclusion was obtained when it was observed that the ESR spectrum changed considerably as B_{12-r} was converted from the "base on" form to the protonated "base off" form (*108*). The pK_a for this equilibrium for B_{12-r} was reported to be 2.5 (*109*). Recent ESR results on corrin(II) derivatives have been obtained with much better resolution of spectra (*110, 111*) (Fig. 17). The characteristic features of the B_{12-r}-spectra are: (A) apparent axial symmetry with $g_\perp = 2.2$ and $g_\parallel = 2.01$; (B) cobalt hyperfine splitting in the $g_\parallel$ band ($I = 7/2$ for ^{59}Co) with $A_\parallel(Co) = 103.6$ G. (C) nitrogen superhyperfine splitting due to the benzimidazole nitrogen for which $A_\parallel(N) = 17.7$ G.

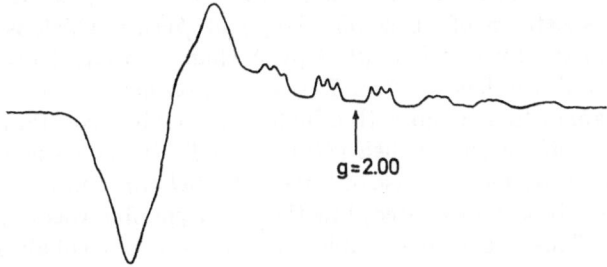

$g = 2.00$

Fig. 17. ESR Spectrum of B_{12-r} taken on a solution of $pH\,7.0$

There is some discrepancy in the literature for the above g values and hyperfine coupling constants, but the above values are representative. The Co(II)-corrins as a group show ESR spectra which can be interpreted on the basis of one unpaired electron occupying an orbital composed largely of the metal $3d_{z^2}$ atomic orbital as is predicted by elementary ligand field considerations. Similar spectra with approximately the same g values, and with resolved cobalt hyperfine in the $g_\parallel$ band have been reported for the closely related complexes of cobalt: (a) Co(II)-phthalocyanine (*112*), (b) Co(II)-porphyrin (*110*) and (c) bis-dimethylglyoximate-Co(II) (*113*). Both *Cockle et al.* (*110*) and *Bayston et al.* (*111*) have shown that for a series of Co(II)-cobinamides, containing different axial bases, $A_\parallel(Co)$ and $A_\parallel(N)$ are inversely proportional. The relationship between these parameters may prove to be very useful when attempting to interpret the ESR spectra of enzyme-bound Co(II)-corrinoids. In addition to the interdependence of the $A_\parallel(Co)$ and the $A_\parallel(N)$, if Co(II) is reduced to Co(I) the base is detached from the 5th coordination position (*111*). The single electron reduction of diaquocobinamide-Co(III) to -Co(II) also gives a five-coordinate derivative. Even in strongly basic solvents containing a nitrogen donor atom, when the water molecule from the fifth coordination position is completely displaced, no hyperfine splitting for nitrogen could be resolved. Piperidine was shown to be the only base which would coordinate under these experimental conditions.

The "base off" form of B_{12-r} therefore shows no nitrogen hyperfine. However, there is some controversy concerning the ESR spectrum of "base off" B_{12-r}. *Schrauzer et al.* (*113*) have reported that "base off" B_{12-r} gives an ESR spectrum similar to that shown in Fig. 17, but with no ^{14}N superhyperfine splitting. *Cockle et al.* (*110*) report a spectrum of "base off" B_{12-r} at pHO which is very different from that reported by *Schrauzer*. The spectrum obtained by *Cockle et al.* is similar to that observed for cobinamide Co(II). *Bayston et al.* (*111*) reported a spectrum of "base off" B_{12-r} at $pH2.1$ which is identical to that reported by *Cockle et al.* $A_{\parallel}(Co)$ changes appreciably in going from "base on" to "base off" for B_{12-r}. Base on has a $A_{\parallel}(Co)$ of 110 G, but this changes to a signal with a high field coupling constant of 156 G for the "base off" species. *Cockle* (*110*) and *Hill* (*41*) have suggested that the spectrum reported as "base off" B_{12-r} by *Schrauzer* in fact does have a nitrogenous base coordinated, but the ^{14}N hyperfine was not resolved adequately. This seems reasonable since *Schrauzer's* cobalt hyperfine coupling constants are too small for "base off" B_{12-r} or cobinamide Co(II).

2. Oxygen Adducts of Co(II) Corrinoids

When molecular oxygen is allowed to react with B_{12-r} (*114*), a $1:1$ complex is formed by this oxidation. Under carefully controlled conditions for this oxygenation, the ESR spectrum changes from that of a typical "base on" B_{12-r} signal (Fig. 17) to a signal where only cobalt hyperfine is observed (Fig. 18). The magnitude of the Co-hyperfine coupling is greatly diminished to a value of 12—18 gauss upon oxygen addition. The greatly diminished value for the Co-hyperfine coupling constant and the disappearance of the nitrogen superhyperfine splitting indicates that the oxygenated complex is a superoxide (*e. g.*, superoxocobalamin, $Co(III)-\dot{O_2^-}$) (*114*).

Since Co(II)-complexes are well known oxygen carriers the formation of O_2-complexes with B_{12}-model compounds has contributed to our understanding of the structure of superoxocobalamin. The reaction of Co(II)-bis(dimethylglyoximato)-pyridine with oxygen gives a diamagnetic dimer in which an oxygen molecule bridges the two cobalt atoms (Co—O—O—Co). Single electron oxidation of this dimer gives a paramagnetic product for which the unpaired electron is delocalized over

Co—O—O—Co. When this μ-superoxocobaloxime is reacted with relatively strong Lewis bases decomposition takes place to give a diamagnetic species and a paramagnetic oxygenated product. (Base → Co—O—O). The ESR spectrum of this oxygenated product is strikingly similar to superoxocobalamin (*115*).

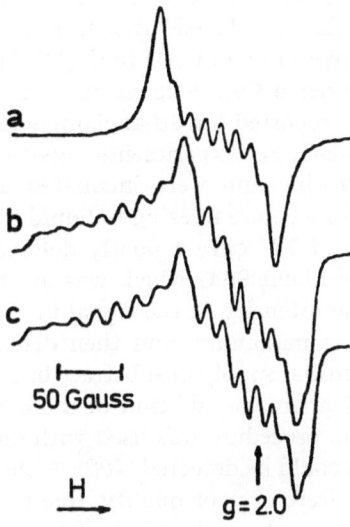

Fig. 18. ESR Spectrum of oxygenated B_{12r} in methanol at 77 K

A similar type of oxygen complex has been observed during the oxidation of $[Co^{II}(CN)_5]^{-3}$ but it was not possible to show that this species was formed in the initial reaction step since with this system, as with the cobaloxime(II) system, the 1:1 adduct apparently reacts very rapidly with another molecule of pentacyanocobaltate(II) to form a diamagnetic binuclear complex with a bridging peroxide ligand (*116*). It appears that in the B_{12}-system the bulk of the corrin ring does not allow formation of the diamagnetic binuclear complex.

3. Application of Co(II) Electron Spin Resonance to the Study of B_{12}-Dependent Enzymes

One of the major reasons for studying the ESR of this series of Co(II)-corrinoids is that this spectral tool may give information on the mechanism of corrin-dependent enzymes. Cleavage of the Co—C—σ-bond has been studied in a few enzymes by using Co(II) ESR, and there is evidence that cleavage of the Co—C-bond is the first step in these enzymatic reactions. In principle the Co—C-bond could be broken in three ways; i. e., homolytically, or heterolytically to either a carbonium ion (CH_3^+) and Cob(I)alamin (B_{12-s}), or a carbanion (CH_3^-) and aquocobalamin (B_{12-a}) (Fig. 6).

If homolysis of the Co—C-bond occurs, then it is expected that a Co(II) ESR signal should appear during enzyme catalysis. Unfortunately, it is possible that a Co(II)-signal could arise from some other side-reaction such as photolysis or oxidation of Co(I) to Co(II). However, it is of interest to determine whether a Co(II)-signal appears during catalysis.

Such a signal was reported for ethanolamine-ammonia-lyase. In the ethanolamine ammonia-lyase experiments, apoenzyme, ethanolamine, and 5'-deoxyadenosylcobalamin were incubated at room temperature for a short period of time before freezing in liquid nitrogen. Examination of the ESR spectrum at 77° gave a poorly defined signal which had a peak-to-peak width of about 90 G which was located at approximately $g = 2.09$. However, the intensity of the signal increased with incubation time passing through a maximum, and then disappearing upon longer incubation. After the initial supply of substrate had been exhausted, then the signal reappeared upon the addition of a second batch of ethanolamine. When the same procedure was used with one of the components missing then no signal could be detected (100). Although the ESR spectra obtained in this work were of poor quality, due to signal to noise problems, obviously this research produced the first evidence for substrate-dependent homolytic cleavage of the Co—C-bond during catalysis.

Ribonucleotide reductase from *Lactobacillus leichmanii* has been studied extensively by Co(II)-ESR. When 5'-deoxyadenosylcobalamin reductase, nucleoside triphosphate, and thiol (reducing agent) were incubated together in the dark, an ESR signal could be observed from the mixture which closely resembled a "base on" cob(II)alamin spectrum (117). All of the components were necessary in order for this signal to be generated. No signal corresponding to an organic free radical was observed. The most striking feature of the ESR spectrum of the enzyme-bound B_{12}-species is the extremely good resolution. While the g values and hyperfine splittings are virtually identical to those in authentic B_{12-r}, the resolution is much higher. Nitrogen hyperfine was even resolved in the highest field line of the $g_{\parallel}$ band and this has never been observed in any spectrum of B_{12-r}-alone.

From these data it seems feasible that a Co(II)-species is generated during catalysis, and that homolysis of the Co—C-bond is a prerequisite for enzyme catalysis in ribonucleotide reductase. However, the kinetics of appearance of the Co(II)-signal indicates that the rate of formation of Co(II) is much slower than either the rate of ribonucleotide reduction or the rate of hydrogen exchange between water and the 5'-methylene group of 5'-deoxyadenosylcobalamin. Therefore, these experiments rule out a Co(II)-species as an obligatory intermediate in this enzymatic reaction. Since a Co(II)-species is formed only under conditions in which ribonucleotide reduction and hydrogen exchange occurs, then it appears that this Co(II)-species arises from a side-reaction of the actual catalytic intermediate. These data indicate that Co(II)-arises from oxidation of Co(I), and that Co(I) is really the active intermediate in ribonucleotide reduction (118).

Bayston et al. (111) have investigated the effect of solvent and added solutes on the degree of resolution of the Cob(II)alamin ESR spectrum. From this study it was concluded that the main effect of solvent and solute is on the position of the benzimidazole ligand with respect to the Co-atom. In a given sample, a lack of homogeneity for the Co—N-bond length is considered a major cause for line broadening in the ESR spectrum. However, the angle of approach by the nitrogen to the cobalt atom may be significant, and this parameter has not been considered.

The very high resolution for the ESR spectrum of cob(II)alamin in the enzyme system is undoubtedly due to the fact that all the coenzyme molecules are bound in an identical environment at the enzyme active site. This results in a homogeneous cobalt-benzimidazole geometry, because both identical binding sites, solvent, and solute molecules can no longer approach the B_{12}-molecule closely. In addition, the enzyme bound cob(II)alamin molecules are more isolated from one another and thus relaxation due to spin-spin interactions is less effective in broadening spectral lines.

A third factor has been suggested for enhanced resolution of the Co(II) ESR spectrum, because in the enzyme the movement of the acetamide and propionamide side-chains of the corrin ring will be restricted. This restriction would diminish fluctuations in the magnetic environment of the cobalt.

The application of ESR to the ribonucleotide reductase system indicates that the catalytic intermediate is a Co(I)-species. The appearance of Cob(II)alamin is probably caused by partial oxidation of the Co(I)-species. In the enzyme bound Co(II)-species the benzimidazole ligand is coordinated, and the corrin ring is bound so tightly that the enzyme produces a unique highly resolved ESR spectrum.

4. Spin Labeling of Vitamin B_{12}

Another potentially very useful method of studying B_{12}-enzymes by electron spin resonance has been developed. This method involves attaching a stable organic free radical, in all cases studied so far a nitroxide, to the B_{12}-molecule and then using the ESR spectrum of the nitroxide to learn something about corrin-enzymes (*119*). Although most of the experiments done in this area are not similar to those of *McConnell et al.* (*120*). the initial work on B_{12} was inspired by some of *McConnell's* work and thus the technique has also been referred to as spin-labeling. In this section we will discuss the chemistry and physical properties of spin-labeled B_{12}-derivatives and in the next section describe how this method has been applied to study one B_{12}-dependent enzyme system.

One obvious way in which to attach a nitroxide group to B_{12} is to simple alkylate Cob(I)alamin with a suitable nitroxide derivative. This would result in having the nitroxide covalently bound to the corrinoid at the upper axial coordination position of the cobalt. Such a procedure is outlined in Fig. 19. In this reaction 4-bromoacetamido 2,2,6,6-tetramethylpiperidine-N-oxyl is used to alkylate Cob(I)alamin. This results in a Co(III)-nitroxalkylcobalamin. The corresponding cobinamide can then be produced by hydrolyzing the ribose-phosphate linkage (*119*).

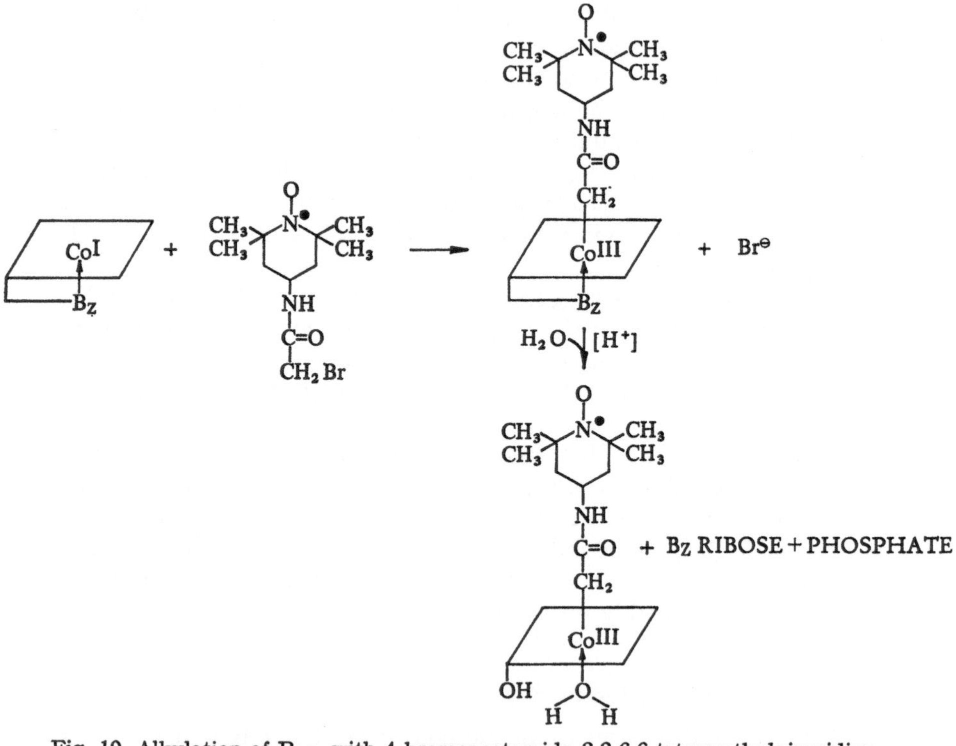

Fig. 19. Alkylation of B_{12s} with 4 bromoacetamido 2,2,6,6-tetramethylpiperidine-N-oxyl to give nitroxalkylcobalamin, and (after hydrolysis) the corresponding cobinamide

Alternatively the cobinamide can be synthesized by first carrying out the hydrolysis of either aquo- or cyano-cobalamin followed by alkylation of the reduced cobinamide.

The nitroxalkylcorrinoids are easy to characterize since they are basically alkyl corrinoids. The u.v.-visible absorption spectra of the nitroxalkylcorrinoids are quite similar to other alkylcorrinoids. The nitroxalkyl cobalamin has a spectrum with λmax at 525, 357, and 329 nm with relative extinctions of 1.1, 1.2, and 0.65 respectively. Spectra for a number of typical alkylcobalamins have been reported by *Firth et al.* *(121)*. The corresponding cobinamide has absorption maxima at 455, 428, 360, and 325 nm with relative extinctions of 1.4, 1.0, 0.61, and 0.62 respectively. Aerobic photolysis of these compounds leaves the corresponding aquocobalamin or aquocobinamide. Likewise, addition of cyanide to the nitroxalkylcobinamide in base leaves dicyano cobinamide.

The electron spin resonance of the nitroxalkylcorrinoids can be readily observed in aqueous solution at room temperature. Both the cobalamin and cobinamide show nitrogen hyperfine coupling constants of 17.2 gauss. A typical spectrum is shown in Fig. 20. The line widths for the low, intermediate, and high field peaks are 1.87, 1.87, and 2.20

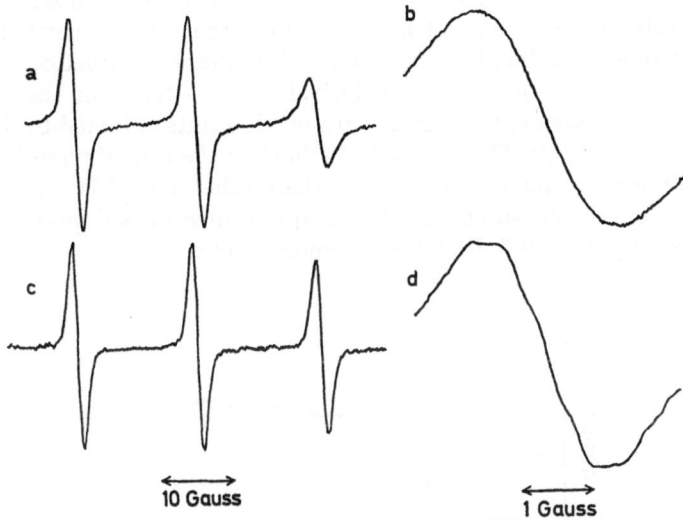

Fig. 20. Electron spin resonance spectra of nitroxalkylcobalamin. (a) Spectrum before photolysis; the high field line is broadened and therefore has a lower peak amplitude. (b) Expanded view of center line before photolysis showing no indication of additional hyperfine from methyl protons. (c) Spectrum of nitroxide photolysis product which has been freed from the cobalamin. (d) Expanded view of center line after photolysis now faintly showing proton hyperfine

gauss, respectively. Broadening of the high field peak also occurs when a free nitroxide is dissolved in a solvent of higher viscosity and is due to incomplete motional averaging of the g and hyperfine tensors. In our nitroxalkylcorrinoids, attachment of the nitroxide to a relatively large molecule (M.W. 1357 for cobalamin) causes the nitroxide to tumble more slowly in solution and this produces the incomplete motional averaging producing the resultant broadening of the high field peak. There is another effect of attaching the nitroxide to a corrinoid which is more subtle. In the free nitroxide there is additional hyperfine coupling to the methyl protons at the 2 and 6 positions of the piperidine ring (Fig. 25). This additional hyperfine splitting is not observable in the corrinoid derivative due to motional broadening.

The differences in the ESR spectra of the coordinated and free nitroxide have been used to study photolysis of the nitroxalkylcorrinoids in a very convenient way. As the nitroxide is released from the corrinoid, the width of the high field line decreases significantly and thus its amplitude increases. The increase in amplitude can be monitored easily by ESR. If one adjusts the spectrometer magnetic field exactly so that one is looking at the maximum of the high field peak and then illuminates the sample while it is in the sample cavity, the increase in intensity can be monitored on the recorder as shown in Fig. 21. Thus it is quite a simple matter to obtain the kinetics of photolysis. Both the nitroxalkylcobalamin and cobinamide exhibited normal first order photolysis kinetics. Under identical conditions the cobalamin photolysis is faster than the cobinamide, and this observation agrees with other results obtained by *Pailes* and *Hogenkamp* (*122*). This method of following the rate of photolysis of alkylcorrinoids is much more direct than other methods which are restricted to the measurement of the appearance of end products of radical chain reactions initiated by homolytic cleavage.

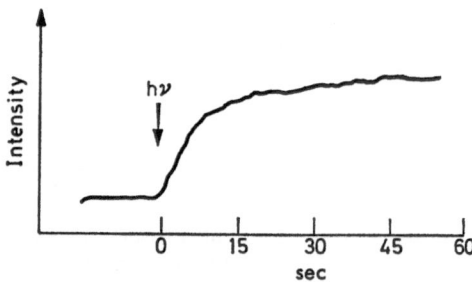

Fig. 21. Photolysis kinetics of nitroxalkylcobalamin. A 10^{-4} M solution was photolyzed aerobically in the spectrometer cavity. The increase in maximum intensity of the high field line is shown as a function of time

Another chemically more interesting spin labeled B_{12} derivative involves coordinate attachment of the nitroxyl function to the cobalt atom of a cobinamide. Fig. 22 shows a reaction in which an alkyl cobinamide is mixed with 4-hydroxy-2,2,6,6-tetramethylpiperidine-N-oxyl. The nitroxide displaces water from the 6th coordination position very slowly and therefore this reaction is usually allowed to proceed for a few days with a large excess of nitroxide. From the properties of the coordinated nitroxide derivative discussed below, it is certain that the cobalt is coordinated by the N—O functional group. An analogous compound to that shown in Fig. 22 can be made with a similar nitroxide in which the 4-hydroxyl-group is missing. In this case the N—O-function is the only basic site on the molecule and therefore must be the position of attachment to the cobalt (119).

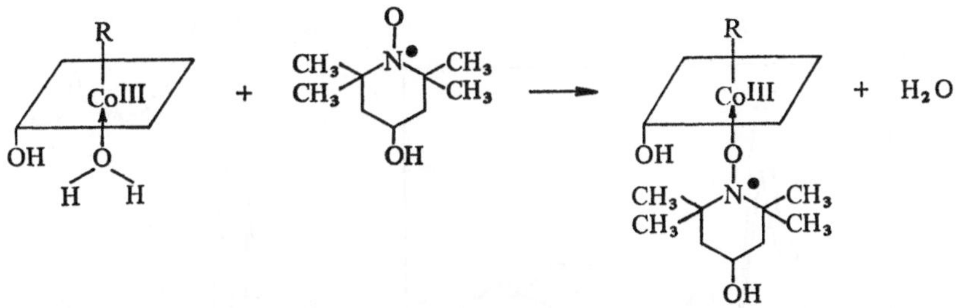

Fig. 22. Synthesis of coordinatively bound nitroxylalkylcobinamide

The u.v.-visible spectrum of the 4-hydroxy-2,2,6,6-tetramethyl-piperidine-N-oxyl-methyl-cobinamide is very similar to methyl-cobinamide itself and as a result this technique cannot be used to rigorously identify the spin labeled derivative. The spin labeled compound does show a spectral change with pH between pH 7.0 and pH 2.0 which methyl-cobinamide does not exhibit. Despite the similarities between methyl-cobinamide and nitroxylmethylcobinamide, the circular dichroism spectrum of the two derivatives are quite different. Fig. 23 shows the marked difference in C. D. spectra of 4-hydroxy-2,2,6,6-tetramethylpiperidine-N-oxyl, methylcobinamide, and a methylcobinamide solution containing an equimolar amount of uncoordinated nitroxide.

Nuclear magnetic resonance studies on spin labeled derivatives are not extremely useful due to the paramagnetism of the molecule. However, the NMR spectrum of spin labeled methylcobinamide confirms that the nitroxyl function is coordinated to the cobalt. It is possible in this compound to obtain good resolution of the methyl group resonance.

This peak is broadened and contact shifted down field by the unpaired electron (Fig. 24). A spectrum of a mixture of methylcobinamide and free nitroxide shows broadening of the methyl resonance but no shift in resonance position. Thus the nitroxide must remain attached to the cobalt atom in solution.

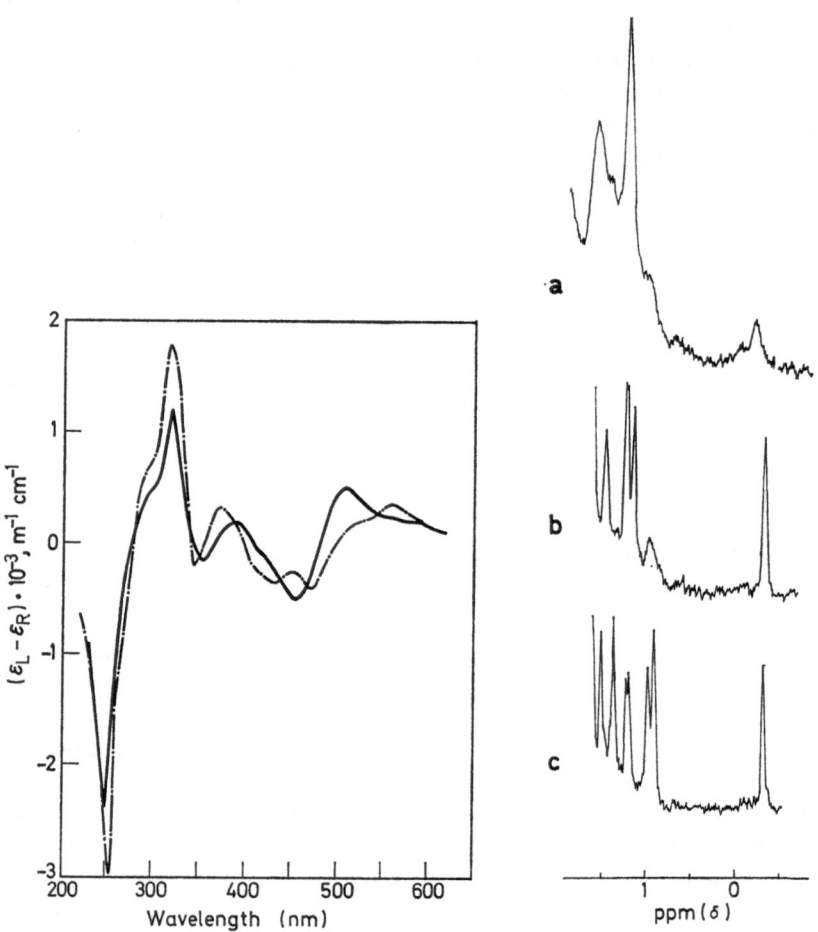

Fig. 23. C. D. Spectra of 4-hydroxy-2,2,6,6-piperidine-N-oxyl-methyl-cobinamide (—) and methyl-cobinamide plus free nitroxide (·—·—) at 3 x 10^{-5} M in ethanol

Fig. 24. 220 MHz NMR spectra of (a) nitroxylmethylcobinamide, (b) methylcobinamide and (c) methylcobalamin at pH 1.2 (base off). In each case the high field peak is the cobalt bound methyl group. In the nitroxyl derivative (a) the methyl group is significantly broadened and shifted downfield

As with the nitroxalkylcobalamins *(119)* and cobinamides, the co-binamides in which nitroxide is coordinated show electron spin resonance spectra very similar to the spectrum of free nitroxide. The high field line is not broadened as much as in the spectrum of a nitroxalkyl-cobinamide. No hyperfine splitting from methyl protons in the 2 or 6 positions can be observed for the bound nitroxide. However, treatment of the coordinate spin labeled compounds with cyanide releases the nitroxide. When this happens, the proton hyperfine can be observed (Fig. 25). Thus treatment with cyanide simply displaces the nitroxide and a spectrum for free nitroxide is observed.

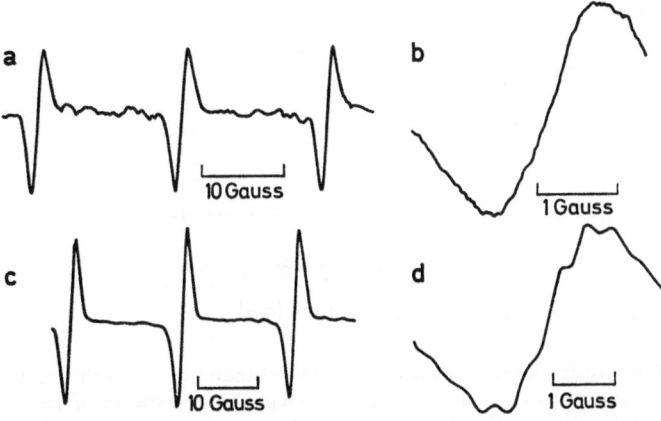

Fig. 25. Electron spin resonance spectra of 4-hydroxy-2,2,6,6-tetramethylpiperidine-N-oxyl aquocobinamide before and after treatment with CN⁻ (a) spectrum of aquo derivative. (b) Expanded view of center line before addition of CN⁻. (c) Spectrum of liberated nitroxide. (d) Expanded view of center line after CN⁻ treatment showing additional proton hyperfine

The photolysis of alkylcobinamides containing coordinatively bound nitroxide is quite interesting. Aerobic photolysis generates an aquocobinamide with the nitroxide still bound in which case the electron spin resonance does not change. Anaerobic photolysis of nitroxylalkyl-cobinamides is interesting to both coordination chemists and biochemists. Anaerobic photolysis of a normal alkyl cobalt corrinoid results in generation of the paramagnetic cobalt(II) species. In the case of coordinated nitroxide derivatives, there are a number of interesting possibilities concerning the nature of the anaerobic photolysis product. Experimentally it is observed that the ESR signal disappears when the sample is photolyzed. If one adjusts the ESR spectrometer to observe the maximum of the signal from one of the hyperfine lines, and then photolyzes the sample anaerobically in the spectrometer sample cavity, then the rate of homolysis by light can be monitored (Fig. 26). Fig. 27 shows the spectrum before and after photolysis. In actual practice the spectrum can be made to completely disappear upon exhaustive photolysis. Not

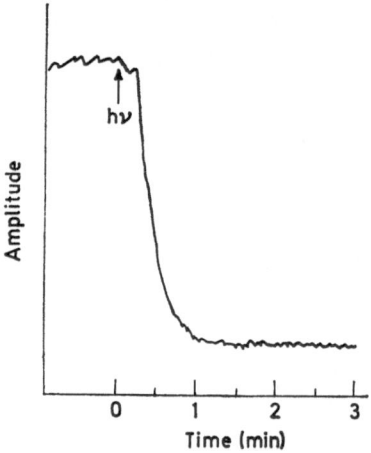

Fig. 26. Photolysis kinetics of nitroxylmethylcobinamide. The solution was photolyzed anaerobically in the spectrometer cavity. Disappearance of the signal was monitored by following the decrease in maximum intensity of high field line as a function of time

only does the original spectrum disappear, but it is impossible to detect an ESR signal at any other resonance position at room temperature after photolysis. The photolysis product must be either diamagnetic, or paramagnetic with the spins coupled in such a way that electron spin relaxation is very rapid so that any ESR signal is broadened enough to escape detection. A diamagnetic intermediate could result from complete pairing of the spins in some molecular orbital composed of appropriate ligand orbitals and the cobalt $3d_{z^2}$ orbital, or from chemical destruction of the nitroxide. The latter possibility is easily ruled out by allowing oxygen to come into contact with the photolyzed sample. When this is done the cobalt(II) is oxidized to cobalt(III) and the original ESR spectrum returns. Furthermore, the nitroxide is still bound to the cobalt. Thus the loss of signal upon anaerobic photolysis is completely reversible. It has not yet been established if the photolysis product is actually diamagnetic or still paramagnetic. The changes in the ESR spectrum described above have been used to great advantage in the study of the enzyme ethanolamine ammonia lyase. Many attempts have been made to generate a cobalt(I) species with bound nitroxide. In this case irreversible loss of the ESR signal occurs due to chemical reduction of the nitroxide by the highly nucleophilic cobalt(I) species.

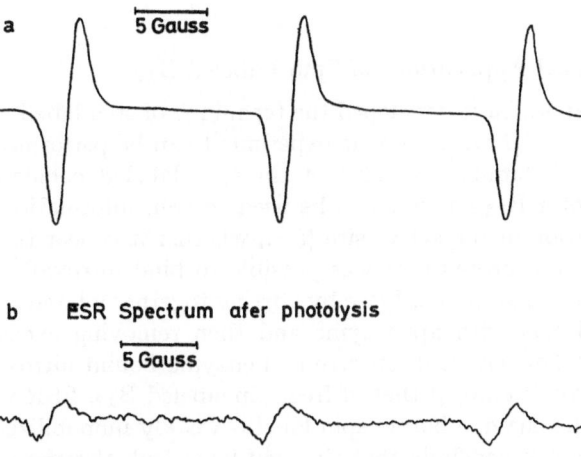

a   5 Gauss

b   ESR Spectrum afer photolysis

5 Gauss

Fig. 27. ESR Spectra of nitroxyl methylcobinamide. (a) Before photolysis and (b) after anaerobic photolysis. In (b) the spectrometer signal level is set much higher than in (a) so that the residual signal intensity is only a few percent of initial intensity

One other aspect of the photolysis of coordinate spin labeled derivatives is of interest. Nitroxides are good free radical scavengers (*123*). As a result, when methyl-cobalamin is photolyzed in the presence of a nitroxide, the methyl radical generated will react with the free nitroxide and cause disappearance of the ESR spectrum (*123*). However, once the nitroxide is coordinated it is no longer susceptible to attack by free radicals. Thus the nitroxyl function is quite well protected from approach by other species.

One of the somewhat surprising aspects of the coordinate nitroxyl-cobinamides is that they show no indication of cobalt hyperfine splitting even though the nitroxyl function is coordinated directly to cobalt $I(^{59}Co) = {}^7/_2$. This is contrary to what has been found in nitroxide adducts with $AlCl_3$ (*124*).

In 1:1 complexes with $AlCl_3$, strong aluminum $I(^{27}Al) = {}^5/_2$ hyperfine coupling has been observed. With di-t-butyl-nitroxide $A_{Al} = 11.40$ gauss. In the coordinate cobinamides any cobalt hyperfine coupling must be less than 0.25 gauss. These data with $AlCl_3$ may be the result of the compound being a much stronger Lewis acid. Also, this discrepancy may arise in part from the fact that the aluminum acceptor orbital in $AlCl_3$ has a large amount of S character, (e. g. sp^3), while the cobalt acceptor orbital is probably a virtually pure $3d_z^2$ orbital with a node at the cobalt nucleus.

5. Biochemical Applications of Spin Labeled B_{12}

McConnell et al., have developed the technique of spin labeling biological macromolecules (*125*). A similar experiment can be performed using the nitroxalkylcobalamins. Binding of the spin labeled cobalamin to the active site of a B_{12} enzyme can be used to gain information about the general location of the active site (*i. e.*, whether it is near the surface or deep within the protein). It was possible to bind nitroxalkylcobalamin to ethanolamine-ammonia-lyase by stirring together a large excess of the spin-labeled B_{12} with apoenzyme and then removing excess free B_{12} by dialysis (*126*). The ESR spectrum of enzyme bound nitroxalkylcobalamin strongly resembled that of free spin labeled B_{12}. ESR was used to show that the enzyme bound spin label is weakly immobilized (*125*). As a result, one can conclude that the spin label and, therefore, the active site is relatively close to the enzyme surface. On the other hand, when a B_{12} binding protein from *Methanobacillus omelianskii* was spin labeled by the same method, the ESR spectrum was very different from unbound spin label. In this protein the ESR data obtained were very similar to a powder spectrum. Therefore, in this protein the spin label is strongly immobilized and it can be concluded that the B_{12} binding site is deep within the protein molecule (*126*).

The most extensive and informative enzyme work with B_{12} spin labels has been carried out on the enzyme ethanolamine ammonia-lyase (*123*). This work has employed six-coordinate 4-hydroxy-2,2,6,6-tetramethylpiperidine-N-oxyl-5′-deoxyadenosylcobinamide. Ethanolamine ammonia lyase uses 5′-deoxyadenosylcobalamin as cofactor. Spin labeled 5′-deoxyadenosylcobinamide was used in order to determine the nature of the first step in the mechanism of action of ethanolamine ammonia lyase by determining the manner in which the Co—C bond is broken. Spin labeled 5′-deoxyadenosylcobinamide was synthesized by taking reduced diaquocobinamide and reacting it with an excess of 5′-tosyladenosine to give 5′-deoxyadenosylcobinamide. This was stirred for three days with a 20 fold excess of 4-hydroxy-2,2,6,6-tetramethylpiperidine to give the spin labeled coenzyme. The chemical and physical properties of the 5'-deoxyadenosyl derivative are similar in most respects to those of the nitroxylmethylcobinamide discussed in the previous section. Details of the absorption spectra are obviously different and the 5'-deoxyadenosyl derivative is more sensitive to visible light. The other difference between the methyl and 5'-deoxyadenosyl spin label is that when methyl is the alkyl ligand, the nitroxide is coordinated more strongly. This complicates the photolysis experiments slightly. When methylcobinamide nitroxide is photolyzed anaerobically and reoxidized, the signal disappears and then returns to its full intensity upon reoxidation. When 5'-deoxyadenosyl is photolyzed anaerobically and reoxidized the signal returns to only about 60% of its original intensity. This is apparently due to the fact that in the 5'-deoxyadenosyl derivative some nitroxide can be displaced which then reacts irreversibly with the free radical generated by homolysis.

There are three possible ways in which the Co—C bond can break. Each of these three mechanisms leaves the cobalt atom in a different formal oxidation state. Therefore, if one monitors the oxidation state of the cobalt during enzyme catalysis, one can infer the mechanism of Co—C bond cleavage. The discussion of the ESR of coordinate nitroxyl cobinamide derivatives in the previous section contains the information necessary to use ESR to follow the cobalt oxidation state. Fig. 28 reviews the possible mechanisms of Co—C bond cleavage. As discussed previously, the ESR spectra of all cobalt(III) derivatives look very similar to the spectrum of free nitroxide. The ESR spectrum disappears when a cobalt(II) species is generated (Fig. 28b), and this loss of signal is reversible since oxygen will oxidize the cobalt back to cobalt(III) and the original ESR spectrum returns (Fig. 28a). Finally, generation of cobalt(I) (Fig. 28c), irreversibly decomposes the free radical and an ESR spectrum cannot be obtained under any conditions.

Spin labeled 5'-deoxyadenosylcobinamide has been used as a cofactor for ethanolamine-ammonia-lyase and the ESR spectrum followed during catalysis (123). This spin labeled coenzyme is biologically active in this enzyme. Enzyme kinetics showed this derivative to have the same V_{max} as the cofactor 5'-deoxyadenosylcobinamide, but it has a higher K_m value of 5.1×10^{-6} M compared to 4.6×10^{-6} for 5'-deoxyadenosylcobinamide (123). When the spin labeled coenzyme was incubated with apoenzyme to give the enzyme-coenzyme complex, the nitroxide ESR spectrum resembled that of free spin label but the lines are broadened significantly.

The addition of ethanolamine to the enzyme-coenzyme complex resulted in greater than 90% loss in signal intensity. When the substrate supply was exhausted, the original ESR spectrum reappeared slowly but did not regain its full intensity after 30 minutes. Under the experimental conditions necessary to carry out the ESR work it was impossible to measure the kinetics for the disappearance of the signal, because the reaction is so rapid. However, when deuterated ethanolamine is used as substrate, it is possible to follow the rate of disappearance of the ESR signal.

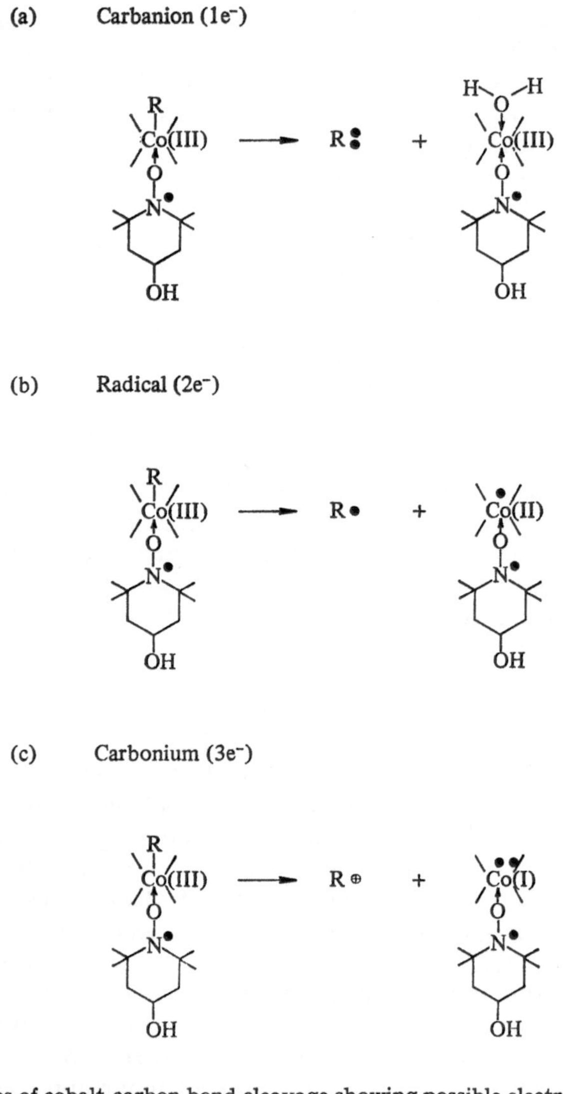

(a)   Carbanion (1e⁻)

(b)   Radical (2e⁻)

(c)   Carbonium (3e⁻)

Fig. 28. Routes of cobalt-carbon bond cleavage showing possible electronic arrangements of the resulting spin labeled derivative

After the enzyme reaction is complete it is possible to remove spin labeled 5′-deoxyadenosylcobinamide from the enzyme by adding another protein which binds corrinoids more strongly than ethanolamine ammonia lyase. When this was done the nitroxide ESR signal returned to its full original intensity. The changes in intensity with time are shown in Fig. 29. Control experiments showed that when ethanolamine was reconstituted with 5′-deoxyadenosylcobalamin and the enzymatic reaction then carried out in the presence of free nitroxide, no changes in the nitroxide ESR could be detected and the presence of nitroxide did not inhibit the reaction. When acetaldehyde and ammonium ions (reaction products) were added to the enzyme-spin labeled coenzyme complex, a slow loss in signal intensity occurred which could be restored by removing the spin label from the enzyme. Thus the conversion of ethanolamine to acetaldehyde and ammonia by the enzyme appears to be partially reversible.

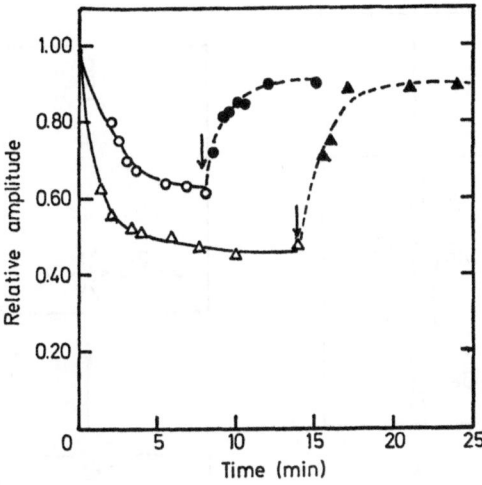

Fig. 29. Decrease in intensity of nitroxide ESR signal upon addition of deuterated ethanolamine to ethanolamine ammonia lyase containing spin labeled cobinamide coenzyme. The two curves are for different concentrations of coenzyme to enzyme. The arrows indicate the point at which alcohol dehydrogenase and NADH was added to remove acetaldehyde from the enzyme. Note that full intensity is regained

The results of this work leads one to the conclusion that the catalytic intermediate is, a cobalt(II) species. The cobalt-carbon bond cleaves homolytically to give an organic free radical. If the initial cleavage gave a carbanion and cobalt(III), the ESR signal would not disappear, and if the cleavage gave a carbonium ion and cobalt(I) the signal would disappear irreversibly. This experiment is therefore in accord with the earlier observation of Babior on the same enzyme that a paramagnetic species is formed during catalysis (*18*). A radical mechanism for the ethanolamine deaminase reaction is proposed in Fig. 30.

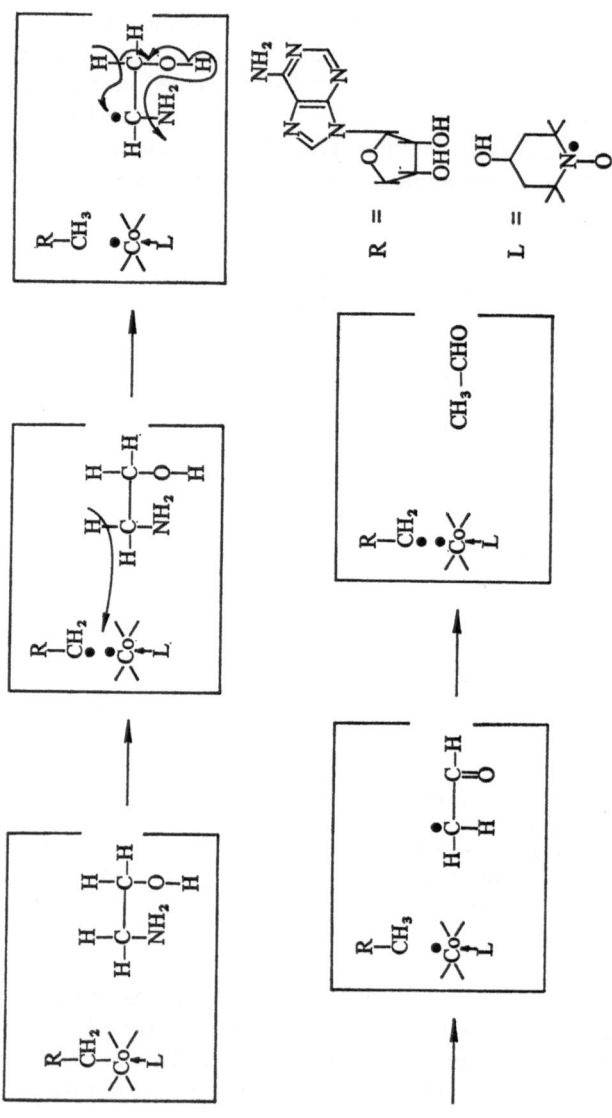

Fig. 30. Proposed free radical mechanism for ethanolamine ammonia lyase

B. Nuclear Magnetic Resonance Spectroscopy of B_{12}-Derivatives

1. Introduction

Most of the NMR work reported on B_{12}-derivatives has been concerned with interpreting spectra and assigning resonance positions. In certain cases some valuable information concerning the chemistry of B_{12} has been obtained. We will discuss the nuclear magnetic resonance work which has been reported for B_{12} plus some of our own unpublished results with particular emphasis on those results which give some insight into vitamin B_{12}-chemistry.

Hill et al. (*127*) reported the first proton magnetic resonance work on a series of cobalamins. This work was carried out at 60 MHz and the spectra are, therefore, of quite low resolution. Subsequently, this work was extended to a wider variety of molecules and also spectra were recorded at 100 MHz (*128*). Five low field resonances and those of the metal alkyl groups were assigned. Some representative chemical shifts for the low field resonances are shown in Table 1.

Table 1 [a), b)]

	Benzimidazole			Ribose	Corrin
	B(2)	B(4)	B(7)	(R1)	C(10)
aquocobalamin	7.25	6.53	6.46	6.28	6.28
hydroxocobalamin	7.12	6.48	6.75	6.29	6.08
vincylcobalamin	7.12	6.42	7.23	6.33	5.95
methylcobalamin	6.97	6.29	7.20	6.29	5.89
ethylcobalamin	7.01	6.26	7.16	6.26	6.02

a) All spectra run in D_2O at pD 7.0.
b) Chemical shifts given in δ (ppm).

In addition to these low field resonances the chemical shift of the coordinated methyl group falls at —0.19 and the methyl portion of the coordinated ethyl group has a chemical shift of —0.61. By comparing these numbers with the chemical shifts of the ethyl protons in the analogous ethyl porphyrin derivative (*129*), in which the resonance positions are all above —5, it has been concluded that there is little or no ring current in the corrin ligand system.

The advent of a commercial 220 MHz NMR instrument has permitted the application of high resolution NMR to many larger molecules and the B_{12}-system is no exception. This technique has allowed more detailed assignments to be made thus permitting the study of more subtle chemical effects, such as changes in molecular conformation.

Fig. 31 shows 220 MHz spectra for methyl cobalamin and methyl cobinamide. Most of the features of the PMR work to be discussed below can be illustrated with these two spectra as references. (See Fig. 1 for references to nomenclature.)

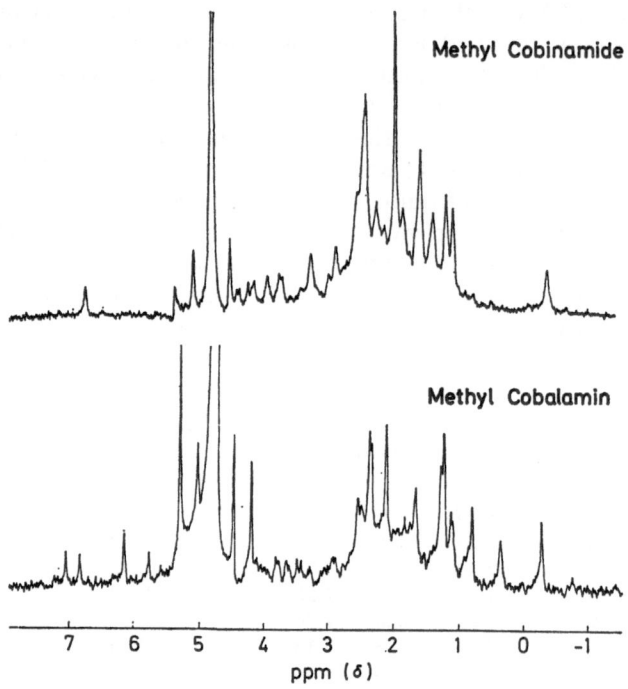

Methyl Cobinamide

Methyl Cobalamin

ppm (δ)

Fig. 31. 220 MHz NMR spectrum of methyl cobalamin in D_2O and 220 MHz NMR spectrum of methyl cobinamide in D_2O

The five low field resonances discussed above and outlined in Table 1 can be readily identified in the region below $\delta = 6.0$ in the methylcobalamin spectrum. When $(CD_3)_2SO$ is used as solvent instead of D_2O, the region shows a complex set of peaks from about 20 protons because all the N—H-protons from the amide side chains are resolved. In D_2O, of course, these protons exchange with solvent deuterons. Notice that in the spectrum of methyl cobinamide only one resonance remains at low field corresponding to the vinyl proton at C(10). The resonance at highest field around $\delta = 0$ is due to the cobalt bound methyl group. This can be easily confirmed by comparing the spectra of methyl and aquo cobalamin. In the latter, the high field resonance is missing even though the rest of the upfield region of the two spectra are quite similar. The high field chemical shift exhibited by cobalt bound alkyl groups could in principle be due to a number of factors such as a small amount of ring current from the corrin ligand, temperature independent paramagnetism of the cobalt, or increased electron density on the alkyl ligand giving it a more carbanion-like character. Probably a combination of these factors is responsible for these observed shifts (130).

In most cobalamin derivatives another peak with the proper intensity for a methyl group appears at approximately 0.5 δ. The position of this peak in the cobalamins is strongly pH dependent and the peak is not apparent in cobinamide derivatives. This interesting resonance can be assigned to the C(20) methyl group at position C(1) of the corrin ring. The results of structural analysis by X-ray crystallography show that when the 5,6-dimethylbenzimidazole is coordinated at the lower axial position, the C(20) methyl group lies approximately 3.2 Å above the plane of the benzimidazole ring system (11—13). As a result this methyl group experiences an upfield shift from the ring current of the aromatic benzimidazole group. This resonance has been used to investigate conformational changes in the molecule as will be described later.

Assignment of the remaining peaks in the cobalamin or cobinamide spectra is less straightforward. The most detailed assignments made thus far have been made by *Brodie* and *Poe* (130). These are outlined in Table 2 for methyl cobalamin.

There are two other interesting features of the PMR spectra of B_{12} derivatives which should be mentioned. In the spectrum of coenzyme-B_{12} where the cobalt bound alkyl group is 5′-deoxyadenosyl, two high field peaks, each with the intensity of one proton, are observed (130). These correspond to the two 5′-methylenic protons.

Table 2. *Some assigned resonance positions in* B_{12}

Resonance position δ (ppm)	Assignment
−.06	cobalt-methyl
−.34	C(20) methyl
1.21	C(47) methyl
2.40	C(5) methyl and C(15) methyl
2.46	
1.07	Pr(3) methyl
3.58	B(10) methyl and B(11) methyl
3.96	
1.74—0.74	C(25), C(36), C(46), C(54) methyls

In the spectra of alkyl cobinamides two peaks have been observed at 3.89 and 4.42 which were assigned to the protons of a water molecule coordinated at the lower axial site (130). To confirm this assignment, it was found that addition of cyanide to methyl cobinamide, which displaces coordinated water, caused the peaks to disappear. Likewise, addition of excess D_2O caused disappearance of the peaks through either ligand exchange or proton-deuteron exchange.

2. Electronic Transmission through the Cobalt Atom (the *cis* and *trans* Effects)

It has been observed by a variety of physical measurements that varying the electronic properties of one of the axial ligands perturbs the electron distribution in both the corrin ring and in the trans axial ligand. These phenomena have been referred to as the *cis* and *trans* effects respectively (131—133). The cis-effect can be demonstrated quite clearly by observing the change in chemical shift of the C(10)-vinyl proton of the corrin ligand. This can be seen from the data in Table 3. It appears that for the cobinamides the more electron withdrawing axial ligands cause the C(10)-resonance to occur at higher field. However, strangely enough, for the cobalamins the reverse-order seems to hold. Thus it appears that the transmission of electronic effects into the cis-corrin ligand is a complicated process in which inductive effects do not necessarily dominate, but rather that more subtle factors such as the conformation of the corrin ring and the stereochemistry of the axial ligands play an important role.

Table 3. *Chemical shift of C (10) vinyl proton in different B_{12}-derivatives (131—133)*

Compound	δ (ppm)
Dicyanocobinamide	5.83
Ethylcobalamin	5.88
Methylcobalamin	5.89
Vinylcobalamin	5.95
Hydroxocobalamin	6.08
Aquocobalamin	6.28
Aquocyanocobinamide	6.5
Methylcobinamide	6.80
Ethylcobinamide	6.95
Isopropylcobinamide	6.96

Despite this fact it is interesting that there is quite a good correlation between the C(10)-chemical shift and the position of the β-band[1]) in the electronic spectrum (*134*). This correlation, shown in Fig. 32, is somewhat unexpected because, as mentioned above, the electronic transmission in the cis-effect seems to be quite complicated.

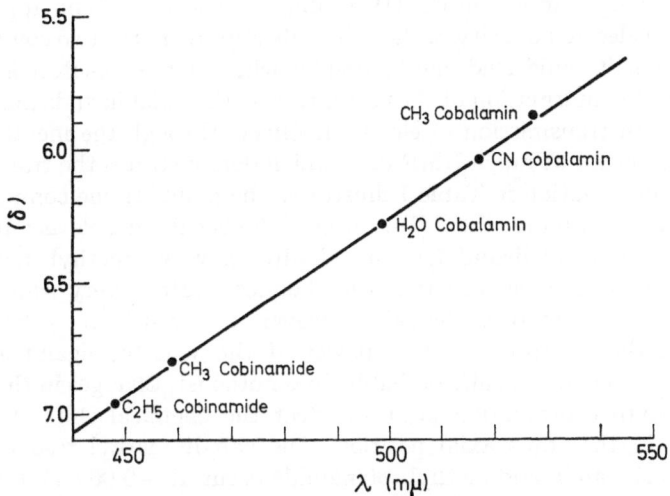

Fig. 32. Correlation between H-10 chemical shift and of the β band in U.V.-visible spectrum

The cis-effect has been studied in a similar manner using cobalt(III)-dimethylglyoxime (dmg) complexes (*135*). These compounds have been proposed as model compounds for B$_{12}$. This system is far less complicated than the cobalamin molecule and offers the possibility of isolating the portion of the cis-effect due solely to inductive electronic transmission. In this work the chemical shift of the ligand methyl groups was investigated as a function of the axial ligand (R) in complexes RCo(dimethylglyoximato)$_2$PPh$_3$. It was found that a linear relationship exists between the chemical shift of the ligand methyl group and the Hammett σ-para function of R for a wide variety of axial ligands.

It is clear from the above work on B$_{12}$ and the inorganic models that changing the axial ligands has a noticeable effect on the corrin system, as evidenced by the C(10)-proton chemical shift. On the other hand,

---

[1]) The β band is the $0 \to 1$ vibrational component of the first electronic transition.

there is currently no evidence that changing the corrin ring, appreciably affects the axial ligands. It is possible to replace the vinyl proton at C(10) with a chlorine atom to give a methyl-10-chlorocobalamin. The electronic spectrum of the chlorinated compound is markedly different from that for the parent methyl cobalamin. If one then compares the chemical shifts of the cobalt methyl protons in methyl cobalamin and methyl 10-chloro-cobalamin they are —0.06 δ and 0.01 δ respectively. Thus even though drastic changes appear in the UV-visible spectra, there is no appreciable change in electron density at the cobalt alkyl group in the two compounds. Such a result could lead one to wonder whether the cis-effect is due to changes in conformation of the corrin ring as the axial ligands are varied, rather than transmission of electronic effects through the metal.

There is not as much NMR data which demonstrates the trans-effect, but the information in Table 1 illustrates the point. If one compares the resonance position for the B(7)-proton of the benzimidazole as a function of the trans axial ligand for aquo, hydroxo, vinyl, methyl, and ethyl cobalamin, it can be seen that the chemical shift it very significantly influenced by the trans ligand. However, a quantitative relationship between the electron donating power of the variable ligand and the B(7)-proton chemical shift probably does not exist. Changes in the ligand at the sixth coordination position affect the chemical shift of methyl protons in the trans axial position. The cobalt methyl resonances in methyl cobalamin and methyl cobinamide occur at —0.06 δ and —0.36 δ respectively. Thus there is a difference of 0.3 ppm in the methyl resonance as the trans ligand is changed from benzimidazole to water. *Brodie* and *Poe (130)* attribute this difference to the effect of the benzimidazole ring current rather than to electronic effects transmitted through the cobalt. More extensive and more readily interpretable experiments on the trans-effect have been carried out by observing changes in the CN stretching frequency in a series of cyanocorrinoids (*133*).

There is a large amount of data available concerning the thermodynamic effects of ligands on other coordination sites (*i. e.*, the thermodynamic *cis*- and *trans*-effects). However, very little is known about the effects of ligands on the kinetic lability of other coordination sites. In fact, very little work has been carried out, directly with B_{12}-derivatives, or with models of B_{12}, on the kinetics of ligand substitution at the cobalt center. Of particular biochemical interest would be studies on the rate of displacement of coordinated benzimidazole by various ligands. Such work has not been reported at present. If the benzimidazole is replaced during enzymatic catalysis so that the lower axial position is occupied by some other Lewis base, one would expect this displacement, and the reverse step, to be very facile. This appears to be qualitatively true in that when water displaces benzimidazole as the benzimidazole is protonated, the NMR spectrum at all pH values shows one resonance position for each type of proton indicating that the "base on" and "base off" species are in rapid enough equilibrium to be in the NMR fast exchange region.

Randall and *Alberty (136)* have studied the binding of various ligands to aquocobalamin using stopped flow techniques. This work suffers from the fact that it is not clear if the added ligand is displacing coordinated water or coordinated benzimidazole. One might be led to believe that the reaction studied in this work is in fact displacement of benzimidazole because the kinetics are at least inconsistent with a mechanism in which unimolecular dissociation of coordinated water is the rate limiting step.

Whatever the detailed mechanism, for the reaction

$$\text{aquo } B_{12} + L \rightleftharpoons \text{aquo } B_{12} - L$$

the following order of association and dissociation rate constants were found as a function of ligands.

Magnitude of association rate constants:

$$SCN^- > N_3^- > CN^- > CNO^- > HN_3 > \text{imidazole}$$

Magnitude of dissociation rate constants:

$$SCN^- > CNO^- > N_3^- > \text{imidazole} > CN^-$$

If, in fact, these data are for ligand replacement of benzimidazole it is of interest from a biochemical standpoint.

It is quite certain that the two cobalamin species with coordinated and free benzimidazole are in rapid equilibrium. However, the ligands bound to the lower coordination site are apparently not always easily displaced. As mentioned above, *Brodie* and *Poe (130)* have found that in DMSO, a water molecule is firmly bound to the sixth coordination position of alkyl cobinamides. Thus, even though DMSO is a good Lewis base, it will not easily displace water from the primary coordination sphere of the cobalt.

Although no systematic work on the kinetics of ligand exchange have been carried out on cobalamins or cobinamides, some interesting work has recently appeared concerning the ligand exchange kinetics in complexes of methyl bis(dimethylglyoxime) cobalt(III). These complexes are the type most often mentioned as inorganic model systems for vitamin B_{12} *(3)*.

Ludwick and *Brown (137)* have investigated the kinetic lability of various ligands (L) in the system $CH_3Co(\text{dimethylglyoximato})_2L$, dmg-dimethylglyoximato anion. Unlike B_{12}-derivates, attempts to remove a ligand from methyl bis(dimethylglyoximato)-Co(III) to produce a 5-co-ordinate species result in the formation of a dimer. This dimer will react with most relatively strong Lewis bases to give back a 6-coordinate mono-meric product. Thus, using this sort of chemistry, a wide variety of com-plexes $CH_3Co(\text{dimethylglyoximato})_2L$ have been prepared. One of the aspect of these complexes which *Ludwick* and *Brown* have investigated involves the rates of exchange between free and bound ligand,

$$CH_3Co(\text{dimethylglyoximato})_2L + L^* \longrightarrow CH_3Co(\text{dimethylglyoximato})_2L^* + L$$

Table 4 shows the NMR coalescence temperatures for this process for a number of ligands. In addition to this data it was shown that the ex-change rates for the ligands acetonitrile, dimethylsulfoxide, and diphenyl-

Table 4. *Coalescence temperatures for several six coordinate methyl cobaloximes*

Ligand	Coalescence Temperature °C
$S(CH_2-CH_2)_2O$ (S-bonded)	55°
$S(CH_3)_2$	65°
$N(CH_3)_3$	75°
$P(C_6H_5)_3$	75°
$P(OCH_3)_3$	105°

sulfoxide are much faster. In fact they show NMR fast exchange at tem-peratures below 0 °C. Thus the order of rates of exchange in this process for a variety of ligands is

$$(C_6H_5)_2SO > CH_3CN \approx (CH_3)_2SO > S(CH_2CH_2)_2O \text{ (S-bonded)}$$
$$> N(CH_3)_3 > P(C_6H_5)_3 > N-C_5H_5 > P(OCH_3)_3 > CNCH_3.$$

From this sort of data it can be concluded that the cobalt(III) ion in this complex is a soft, or class *b*, Lewis acid. It has been suggested pre-viously that the cobalt(III) ion in vitamin B_{12} was also a class *b* acid (*138*). The kinetic order, which presumably corresponds closely to the order of thermodynamic stability, suggests that for at least some ligands π-bonding may be important. In this connection it is noteworthy that methylcobaloxime forms a complex with CO (*139*).

The ligand exchange kinetics between free and coordinated ligands were studied more thoroughly for the ligands trimethylphosphite and triphenylphosphine. The kinetics were analyzed using conventional NMR line shape technique (140).

It was found that the exchange rate was independent of the amount of excess free ligand. As a result it appears that the rate determining step in this process is dissociation of the 6-coordinate complex to give an intermediate which recombines very rapidly with free ligand. By studying the temperature dependence of the rate of exchange, activation energies of 23.3 ± 0.3 and 20.5 ± 1.2 kcal/mole were calculated for trimethyl phosphite and triphenyl phosphine complexes respectively.

A comparison of the kinetic behavior of the axial positions in cobaloximes and B$_{12}$-derivatives shows substantial differences. The exchange kinetics of the ligands investigated show much slower rates than what one finds in biochemical processes. Thus, such a slow step would not be found in an enzymatic reaction. Furthermore, the fact that the cobaloximes dimerize, complicates any investigation of ligand exchange kinetics. B$_{12}$, on the other hand, can exist in a stable 5-coordinate form. The results of such differences between B$_{12}$ and cobaloximes leads one to question whether cobaloximes are appropriate model systems with which to study kinetic aspects related to the mechanism of catalysis of B$_{12}$. The corrin ring surely gives the cobalt ion many of its unique properties in B$_{12}$ like molecules and perhaps model studies should focus on cobalt complexes containing macrocyclic rings more closely related to the corrin ring.

3. Conformational Studies by NMR

Earlier it was mentioned that the peak occurring at about 0.5 δ was due to the C(20)-methyl group shifted to higher field by the benzimidazole ring current. The position of this peak is quite pH-dependent. It is well known that benzimidazole can be displaced by water below the pH for protonation of this base. The pK_a-values for this displacement are known for various B$_{12}$-derivatives (122). When the benzimidazole nitrogen is protonated, it is displaced from the primary coordination sphere of the cobalt. As a result, the 5,6-dimethylbenzimidazole group undoubtedly swings free of the corrin ring no longer shielding the C(20) methyl group. Thus when the cobalamin molecule is in the "base off" form the C(20) methyl resonance position returns to a normal value of 1—2 δ.

When the pD of a D$_2$O solution of methyl cobalamin is lowered, the resonance position for the C(20)-methyl shifts progressively to lower field and the intensity of the peak remains constant (50). Thus rather than observing two C(20)-methyl resonances corresponding to the base-on and base-off species, one finds a single peak at a position corresponding to a mole fraction weighted average of the natural resonance positions of the base-on and base-off molecules (140). It is apparent, therefore, that the rate of exchange of the benzimidazole group between the bound and free environment is in the NMR fast exchange limit. The resonance position of the C(20)-methyl group as a function of pD is given in Table 5.

Table 5. pD *dependence of* C (20) *methyl chemical shift*[a])

pD	Chemical shift (δ)
4.38	0.43
3.84	0.46
2.94	0.70
2.80	0.77
2.68	0.83
2.54	0.88
2.41	0.91
1.73	0.97

[a]) The chemical shifts were obtained at 220 MHz with D_2O solutions of 0.02 M methyl cobalamin. The pD was adjusted with D_2O solutions of NaOD and DCl. The pD of each solution was measured just prior to running the spectrum. After the spectrum had been run, the pD of each sample was adjusted to 7 and the spectrum re-run. All spectra were obtained at ambient temperature. DSS was used as an internal standard.

When the data in this table are plotted, the graph shown in Fig. 33 is obtained. From this one can calculate a pK_a of 2.85 for displacement of benzimidazole in D_2O. In addition, since room temperature is above the coalescence temperature, it is possible to set a lower limit on the exchange rate between coordinated and uncoordinated benzimidazole of 3.1×10^2 sec^{-1}. From Fig. 33 one can, by extrapolation, calculate the C(20)-methyl resonance of the "base-on" and "base-off" forms to be 0.41 and 1.05 respectively. These numbers can be used, with the assumption of fast exchange, to determine the relative amounts of "base-on" and "base-off" species from the observed C(20)-chemical shift for any arbitrary sample. Such information would be useful, for instance, when investigating the displacement of benzimidazole by other Lewis bases. Thus for the simple case of benzimidazole displacement we have shown that NMR provides a method for studying the molecular conformation of vitamin B_{12}.

A difference in the solid state and solution conformation of 5'-deoxyadenosylcobinamide (cobinamide coenzyme) has been inferred from the NMR spectrum of cobinamide coenzyme (123) which indicates that the resonance due to the proton at carbon R-1 of the ribose is shifted significantly upfield from its expected position. Such an upfield shift would probably have to arise from the ring current of the adenine ring.

In the solid state structure the adenine ring lies parallel to the corrin ring and in this conformation the R(1)-H is not in a suitable position to experience an upfield shift from the adenine (123). However, a three dimensional model of the cobinamide coenzyme indicates that when the adenine ring lies perpendicular to the corrin, the R(1)-H is in a position to experience an upfield shift arising from the adenine ring current. Thus these data provide experimental evidence for the movement of the adenine from a horizontal to vertical position in solution. This conformational change seems to "uncover" the 5'-methylene protons making the Co—C-bond more accessible to other molecules (123). In addition to conformational changes of the axial ligands, evidence is available for isomerization reactions of the corrin ring.

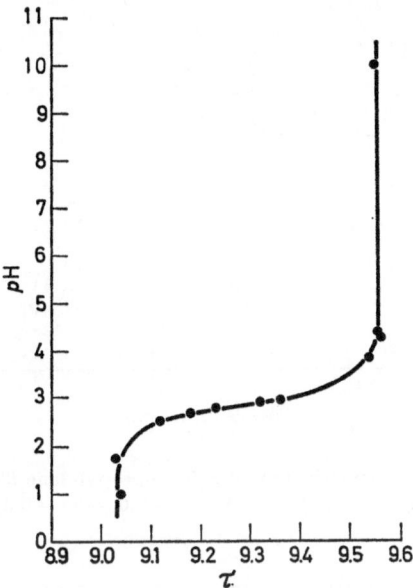

Fig. 33. Change in C(1)-methyl resonance position in methylcobalamin as a function of pH

Hill et al. (141) first observed that light-stable yellow alkylcorrinoids could be synthesized. When methylcobalamin was treated with picrate, a yellow corrinoid was isolated which was shown to be stable to light, but unstable to cyanide. Similarly, *Taylor* and *Weissbach* (67) demonstrated that the methylcorrinoid-enzyme complex of N^5-methyl-tetrahydrofolate-homocysteine transmethylase was stable to light, but isolation of this methylcorrinoid from the enzyme yielded photolabile methylcobalamin. The red to yellow transition in the cobalamins was attributed to either protonation of the corrin ring at C_{10}, or an inability of the 5,6-dimethylbenzimidazole moiety to coordinate to the cobalt atom.

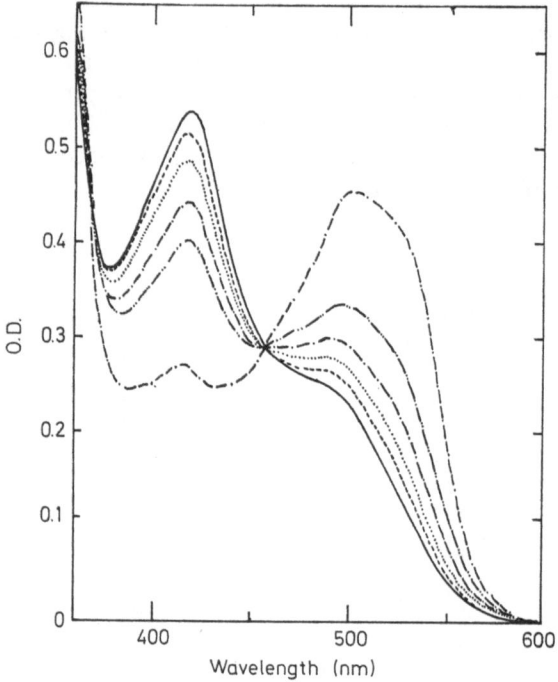

Fig. 34. pH dependent isomerization of 2′,3′-isopropylidene 5′-deoxy-β(D)-ribosyl-cobinamide at pH — 7.0, ———8.0, $\cdots$ 8.5, — $\cdot$ — 9.0, — $\cdot\cdot$ — 9.5, and —— $\cdot$ —— 11.0

Recently, *Law* and *Wood* (*142*) synthesized 2′,3′-isopropylidene 5′-deoxy-β(D)-ribosylcobinamide and they showed that this compound could exist as two different corrin ring isomers. Isomerization was shown to be pH dependent and solvent dependent (Fig. 34). The isomer (Y) with λ_{max} at 420 nm is stable to visible light, but the isomer (R) with λ_{max} at 475 nm is photolabile. A 220 MHz study of these two isomers gave markedly different spectra. For (R) the C_{10} vinyl proton is located at 6.7 δ, but this is shifted to 7.48 δ in the (Y) form. This Y to R isomerization is reversible, and clearly represents a significant conformational change of the corrin ring. Corrin ring isomerizations may be very important for the interaction of corrinoids with proteins.

4. ^{19}F NMR Studies on Fluoroalkylcobalamins

It is possible to synthesize alkylcobalamins containing halogenated alkyl groups. Of these, the compounds studied most extensively are the halo-methylcobalamins (143). For the various fluorine containing derivatives which have been prepared, ^{19}F NMR spectra have been obtained and the results are shown in Table 6. Chemical shifts of a number of fluorome-thanes are included for comparison.

Table 6. ^{19}F-*Chemical shifts in a series of fluoro-methylcobalamins and fluoromethanes*

Compound	Chemical shift, (ppm upfield from $CFCl_3$)
$CFCl_3$	0
CF_2Cl_2	16.3
trifluoromethylcobaloxime	30.6
CF_3Cl	39.9
CF_4	76.7
trifluoromethylcobalamin	94.2
difluorochloromethylcobalamin	94.2
CF_3H	94.9
CF_2H_2	156.7
difluoromethylcobalamin	212.0
fluorodichloromethylcobalamin	213.5
CFH_3	286.7

Fluorine chemical shifts are usually considered to be governed by the local paramagnetic contributions to the shielding constant (144). Thus it has been pointed out that nearly all fluorine containing compounds show ^{19}F-chemical shifts downfield from HF, because in HF the fluorine is nearly ionic with essentially no local paramagnetic contribution to the shielding constant. On the other hand, since in F_2 there is a covalent bond between the atoms, each fluorine atom deviates greatly from having a spherical electron cloud so there should be a large paramagnetic contribution to the shielding, and experimentally F_2 is shown to have a chemical shift far downfield from most other fluorine containing compounds. As a result it is not unreasonable that ^{19}F-chemical shifts have been interpreted in terms of the amount of covalent character in the bond to the fluorine. This has resulted in correlations being made between ^{19}F-chemical shifts and the electronegativity of the group bound to the fluorine. Thus in the series CF_4, CF_3H, CF_2H_2, CH_3F (see Table 6), the electronegativity of the groups attached to a fluorine atom change as

$$CF_3 > CF_2H > CFH_2 > CH_3$$

and the chemical shifts for these compounds increase for the above series reflecting more ionic character in the C—F-bond as one proceeds from CF_4 to CH_3F. Complications arise in the fluorochloromethanes, in which the ^{19}F-chemical shifts come at lower fields than expected, even below the chemical shift for CF_4. This has been explained by saying that in these compounds, the fluorine can more readily undergo π bonding to carbon (i. e., resonance form (I) below is favored over (II)). This then results in less ionic character in the C—F-bond and a larger paramagnetic contribution to the shielding constant.

(I)   (II)

With respect to the cobalamins, the data in Table 6 shows that the chemical shift of CF_3-cobalamin and CF_3H are nearly identical and CF_2H-cobalamin shows a chemical shift midway between CF_2H_2 and CFH_3. Thus one might say quantitatively that the cobalamin is about as electron withdrawing as a hydrogen atom. Likewise, the chemical shift for CF_2Cl-cobalamin is reasonable with respect to the chemical shifts of the substituted methanes. However, $CFCl_2$-cobalamin shows a chemical shift to much higher field than expected and is clearly anomalous. This can be explained by noting that the bulk of the two chlorines is great enough to cause severe steric strain in the cobalt-carbon bond. (This contention is easily confirmed by inspecting appropriate space filling molecular models.) If the Co—C-bond were artificially lengthened to place an abnormal amount of electron density on the alkyl group, then the carbon atom would be less electronegative which would allow a more nearly spherical distribution of electron density about the fluorine atoms resulting in the high field chemical shift. Thus, the ^{19}F NMR is consistent with an artificially lengthened Co—C-bond producing a $CFCl_2$-group with more carbanion character than normal (90). The photolysis rates of a series of halomethylalkyl cobalamins show a correlation with the electronegativities of the alkyl group. However, $CFCl_2$-cobalamin shows an abnormally rapid photolysis rate indicating a weak Co—C-bond. Like-wise, there is a correlation between the alkyl group electronegativity and the pK_a of displacement of benzimidazole for a number of cobal-amins. Once again the $CFCl_2$-derivative is anomalous, showing a lower pK_a than expected. This stronger interaction with the cobalt indicates a more positively charged metal ion which would necessarily result from an increase in electron density on the alkyl ligand. Fig. 35 shows the above correlations in graphical form. Thus the ^{19}F NMR on fluoromethyl cobalamins can be interpreted in much the same way as in fluoromethanes except that $CFCl_2$-cobalamin appears to have a significant amount of strain associated with the Co—C bond which causes the ^{19}F-chemical shift to be anomalous.

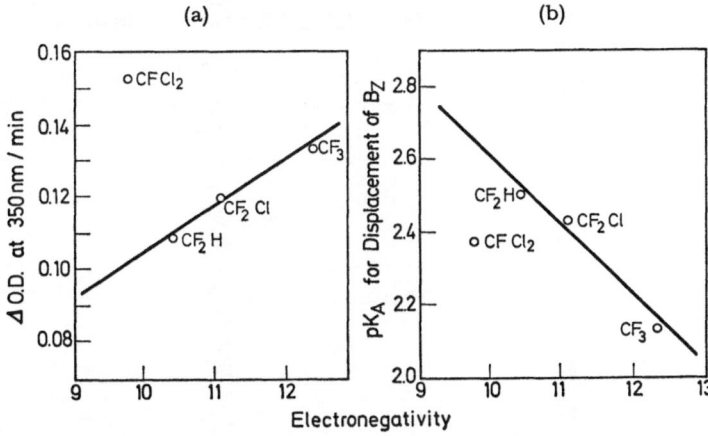

Fig. 35. (a) Correlation between the photolysis rate of halomethyl-cobalamins and the electronegativity of the alkyl ligand. (The electronegativity values are shown as the sums of the Pauling electronegatives of the atoms attached to the methyl carbon.) (b) Correlation between pK_a for displacement of benzimidazole and alkyl group electronegativity for a series of halomethyl-cobalamins

One last point should be mentioned with respect to the ^{19}F NMR results. Included in Table 6 is the chemical shift for CF_3-cobaloxime. The chemical shift for this species is at significantly lower field than that for CF_3-cobalamin. It therefore, appears that the bis(dimethylglyoximato) ring system is considerably more electron withdrawing than the corrin system. As a result, the Co—C-bond is considerably more stable in these systems. These data once again lead one to question to what extent these inorganic systems should be considered useful models for B_{12}.

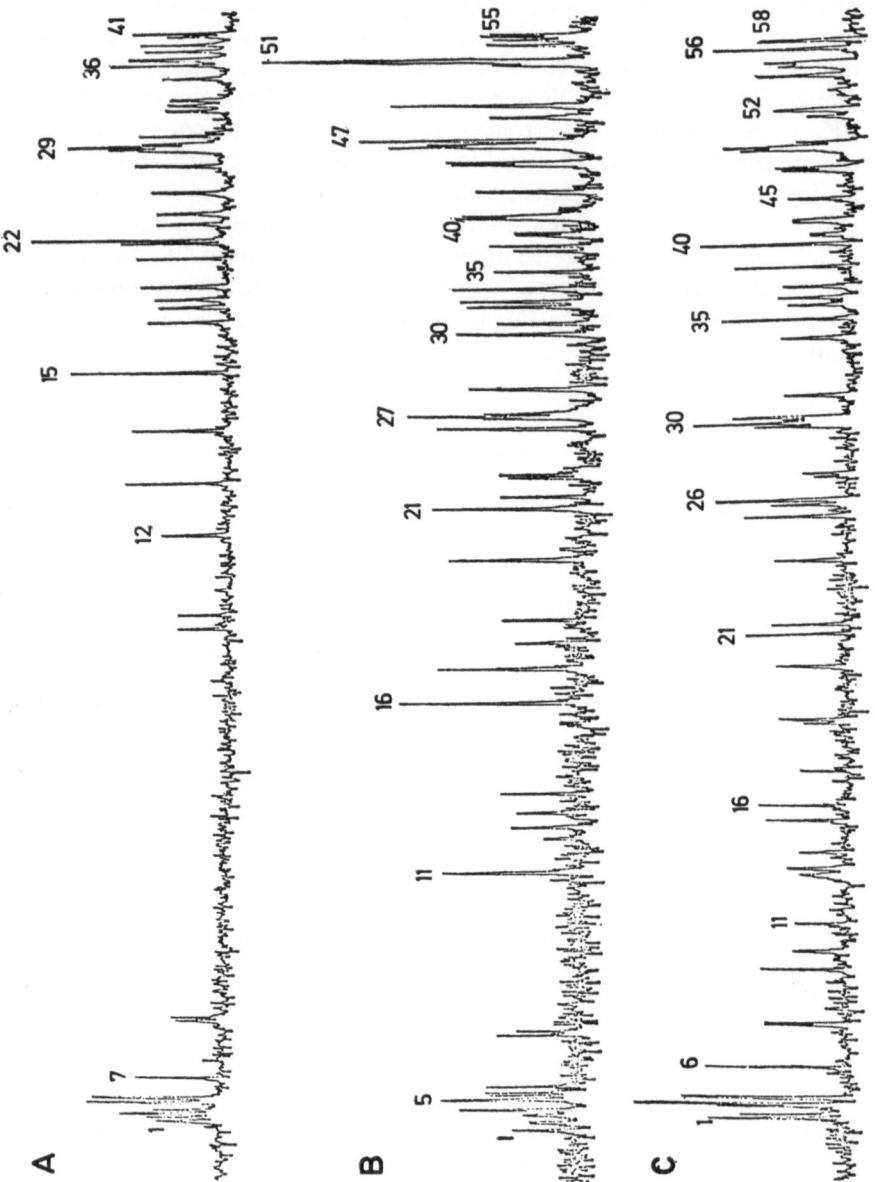

Fig. 36. Proton-decoupled natural-abundance carbon-13 NMR spectra of some corrinoids at 15.08 MHz, obtained by the Fourier transform method. (a) 0.67 M aqueous dicyano-cobinamide. (b) 0.024 M aqueous cyanocobalamin. (c) 0.038 M 5′-deoxyadenosylcobalamin (compliments of *A. Allerhand*)

5. The Use of Carbon-13 NMR to Study B_{12}-Derivatives

One of the most promising applications of nuclear magnetic resonance to vitamin B_{12}-chemistry is the use of carbon-13 NMR (145). The use of ^{13}C fourier transform NMR has greatly increased the applicability of ^{13}C NMR in that it permits one to obtain high quality spectra with natural abundance [13]C.

Doddrell and *Allerhand* have obtained the ^{13}C spectra of a number of B_{12}-derivatives using a Fourier transform instrument operating at 15.08 MHz (145). They have reported spectra for dicyanocobinamide, cyanocobalamin, dicyanocobalamin, and 5'-deoxyadenosylcobalamin. The samples all contained natural abundance ^{13}C and were obtained in concentrations as low as 0.024 M. Representative spectra are shown in Fig. 36. A great advantage of using carbon-13 is that the variation in chemical shifts among nuclei in different chemical environments is much larger than for protons. Therefore, the individual peaks do not overlap appreciably as they do in even the 220 MHz proton spectra. By using several kinds of information it has been possible to assign many of the resonances to individual carbon atoms. The way in which these assignments are made includes 1) comparison of chemical shifts to small model compounds such as benzimidazole; 2) comparison of chemical shifts among the various derivatives; 3) off resonance single-frequency decoupling (145); 4) spin-spin coupling to the ^{31}P-nucleus; 5) partially relaxed spectra, and 6) ^{13}C-spin-lattice relaxation times (145). Details of the assignments and the chemical shift values assigned to the various nuclei are discussed fully by *Doddrell* and *Allerhand* (145). *Doddrell* and *Allerhand* also have shown how powerful carbon-13 NMR may be for the study of conformational changes in the corrinoids. Aquocyanocobyric acid is well known to exist in two isomeric forms (146). The two isomers correspond to two possible coordination isomers involving the axial ligands. Isomerization is thought to proceed through a dicyano intermediate

$$H_2O-Co-CN \xrightarrow{\text{CN}^{\ominus}} CN-Co-CN \xrightarrow{\text{H}_2\text{O}} CN-Co-H_2O$$

Previously proton NMR was used to show that both isomers existed in solution (146). This was due to the fact that two slightly different resonance positions were observed for the C-10-proton. Other than that no evidence for two species was available from the NMR spectrum. On the other hand, the ^{13}C-spectrum shows significant differences in the chemical shifts of many nuclei in the two isomers. Fig. 37 compares the ^{13}C-spectrum of aquo cyanocobyric acid with that for the dicyano derivative in which there is only one isomer. In the dicyano compound there are 38 resolvable peaks while there are 60 in the aquocyano deriva-

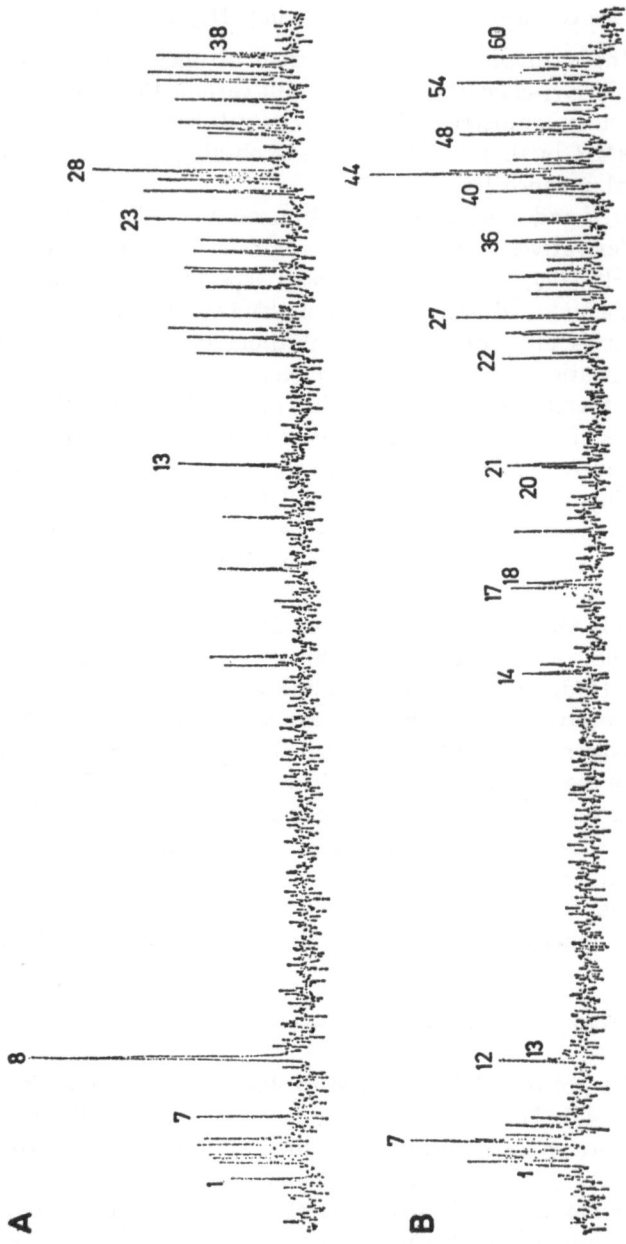

Fig. 37. Proton-decoupled natural-abundance carbon-13 NMR spectra. Peaks are numbered from left to right. (a) 0.064 M aqueous dicyano-cobyric acid. (b) 0.064 M aqueous aquocyanocobric acid. (compliments of *A. Allerhand*)

tive. ^{13}C NMR is far superior to proton NMR for determining the presence of the two similar isomers. The difference in the two isomers are reflected in all parts of the spectrum including even the carbon atoms on the propionamide and acetamide side chains. Thus it is likely that a significant conformational change occurs in the corrin ring upon passing from one isomer to the other. The nature of this conformational change is unknown. In B_{12}, the corrin ring is quite flexible. As the metal ion is pulled away from the ring the corrin ligand can adjust to maximize the bonding interaction with the cobalt. Small adjustments of this sort would produce significant movement of the side chains and thus could account for the observation of differences in the isomers in all regions of the ^{13}C-spectrum.

The most exciting research is yet to be performed on ^{13}C NMR of the corrin enzymes. This could be accomplished by biosynthesis of C-13 enriched samples of biologically active B_{12} derivatives followed by their incorporation into enzymes. Since it has been shown that ^{13}C-spectra of corrinoids are well resolved, and sensitive to small changes in the molecular conformation, then one could hope to get quite detailed information pertinent to the binding of B_{12} and to the mechnism of enzyme catalysis.

Acknowledgements. The authors are indebted to the research efforts of their colleagues Dr. *F. Scott Kennedy*, Dr. *T. Buckman*, Dr. *R. E. DeSimone*, Dr. *B. M. Babior*, Dr. *M. W. Penley* and Dr. *R. S. Wolfe*. Some of the research reported in this review was accomplished by graduate students *P. Y. Law*, *E. L. Lien*, *B. C. McBride* and *F. Shipway*.

VIII. Concluding Remarks

The application of magnetic resonance techniques to biological systems is a relatively new approach for the study of macromolecules. In this review we have presented the different approaches which have been made to study B_{12}-enzymes. Clearly some progress has been made particularly from the application of ESR to a study of the enzymes ethanolamine ammonia-lyase and ribonucleotide reductase. Although ^{13}C NMR is well in its developmental stages it is obvious that this technique will prove to be very useful for the examination of coenzyme-enzyme interactions. Studies of how corrinoids bind in enzymes and how sulfhydryl containing proteins are involved in enzyme catalysis comprise two major problems which must be overcome before realistic mechanisms can be presented for this group of enzymes.

IX. References

1. *Shemin, D., Bray, R. C.:* Ann. N. Y. Acad. Sci. *112*, 615 (1964).
2. *Friedman, H. C., Cagen, L. M.:* Ann. Rev. Microbiol. *24*, 159 (1970).
3. *Schrauzer, G. N.:* Accounts Chem. Res. *1*, 97 (1968).
4. *Minot, G. R., Murphy, W. P.:* J. Am. Med. Assoc. *87*, 470 (1926).
5. *Smith, E. L.:* Nature *161*, 638 (1948).
6. *Rickes, E. L., Brink, N. G., Koniuszy, F. R., Wood, T. R., Folkers, K.:* Science *107*, 396 (1948).
7. *Brink, N. G., Folkers, K.:* J. Am. Chem. Soc. *71*, 2951 (1949).
8. *Kacyka, E. A., Heyl, D., Jones, W. J., Folkers, K.:* J. Am. Chem. Soc. *74*, 5549 (1952).
9. *Wolfe, D. E., Jones, W. H., Valiant, J., Folkers, K.:* J. Am. Chem. Soc. *72*, 2820 (1950).
10. *Hodgkin, D. C., Kamper, J., MacKay, M., Pickworth, J., Trueblood, K. N., White, J. G.:* Nature *178*, 64 (1956).
11. *— — Lindsey, M., MacKay, J., Pickworth, J., Robertson, J. H., Shoemaker, C. B., White, J. G., Rosen, R. J., Trueblood, K. N.:* Proc. Roy. Soc. (London), *A 242*, 228 (1957).
12. *— Lindsey, J., MacKay, M., Trueblood, K. N.:* Proc. Roy. Soc. (London), *A 266*, 475 (1962).
13. *— — Sparks, R. A., Trueblood, K. N., White, J. G.:* Proc. Roy. Soc. (London), *A 266*, 494 (1962).
14. *Lenhert, P. G., Hodgkin, D. C.:* Nature *192*, 937 (1961).
15. *White, J. Y.:* Proc. Roy. Soc. (London), *A 266*, 440 (1962).
16. *Das, P. K., Hill, H. A. O., Pratt, J. M., Williams, R. J. P.:* Biochim. Biophys. Acta *141*, 644 (1967).
17. *— — — —* J. Chem. Soc. (A), 1261 (1968).
18. *Jaselkis, B., Diehl, H.:* J. Am. Chem. Soc. *76*, 4345 (1954).
19. *Tackett, S. L., Collatt, J. W., Abbott, J. C.:* Biochemistry *2*, 919 (1963).
20. *Barker, H. A., Weissbach, H., Smyth, R. D.:* Proc. Natl. Acad. Sci. U. S. *44*, 1093 (1958).
21. *Barker, H. A.:* Federation Proc. *20*, 956 (1961).
22. *Smith, E. L., Mervyn, L., Johnson, A. W., Shaw, N.:* Nature *194*, 1175 (1962).
23. *Bernhauer, K., Muller, O., Muller, Y.:* Biochem. Z. *336*, 102 (1962).
24. *Volcani, B. E., Toohey, J. I., Barker, H. A.:* Arch. Biochem. Biophys. *92*, 381 (1961).
25. *Toohey, J. I., Barker, H. A.:* J. Biol. Chem. *236*, 560 (1961).
26. *Lindstrand, K.:* Nature *204*, 188 (1964).
27. *— Stahlberg, K. G.:* Acta Med. Scand. *174*, 665 (1965).
28. *Irion, E., Ljungdahl, L.:* Biochemistry *4*, 2780 (1965).
29. *Hogenkamp, H. P. C.:* Ann. Rev. Biochem. *37*, 225 (1968).
30. *Stadtman, T. C.:* Science *171*, 859 (1971).
31. *Barker, H. A.:* The Bacteria (I. C. Gunsalus and R. Y. Stonier, eds.) Vol. 2, Chapter 3. New York: Academic Press 1961.
32. *Wood, H. G., Kellermeyer, R. W., Stjernholm, R., Allen, S. H. G.:* Ann. N. Y. Acad. Sci. *112*, 661 (1964).
33. *Abeles, R. H., Lee, H. A.:* J. Biol. Chem. *236*, 2347 (1961).
34. *Smiley, K. L., Sobolov, M.:* Arch. Biochem. Biophys. *97*, 538 (1962).
35. *Bradbeer, C.:* J. Biol. Chem. *240*, 4675 (1965).
36. *Stadtman, T. C.:* Ann. N. Y. Acad. Sci. *112*, 728 (1964).
37. *Dinning, J. S.:* J. Am. Chem. Soc. *81*, 3804 (1959).

38. *Guest, J. R., Friedman, S., Woods, D. D.:* Nature *195*, 340 (1962).
39. *Blaylock, B. A., Stadtman, T. C.:* Ann. N. Y. Acad. Sci. *112*, 799 (1964).
40. *Ljungdahl, L., Irion, E., Wood, H. G.:* Biochemistry *4*, 2771 (1965).
41. *Hill, H. A. O.:* Coordination Chemistry (Editor *G. Eichorn*, 1971) (in press).
42. *Hogenkamp, H. P. C.:* Ann. N. Y. Acad. Sci. *112*, 552 (1964).
43. — Biochemistry *5*, 417 (1965).
44. — *Rush, J. E., Swenson, C. A.:* J. Biol. Chem. *240*, 3641 (1965).
45. *Schrauzer, G. N., Sibert, J. W., Windgassen, R. J.:* J. Am. Chem. Soc. *90*, 6681 (1968).
46. *Wood, J. M., Kennedy, F. S., Rosen, C. G.:* Nature *220*, 173 (1968).
47. *Hill, H. A. O., Pratt, J. M., Ridsdale, S., Williams, F. R., Williams, R. J. P.:* Chem. Commun. 341 (1970).
48. *Schrauzer, G. N., Weber, J. H., Beckham, T. M., Ho, R. K. Y.:* Tetrahedron Letters *3*, 275 (1971).
49. *Wood, J. M., Penley, M. W., DeSimone, R. E.:* Mercury Handbook, I. A. E. C. (1971) (in press).
50. *Penley, M. W., DeSimone, R. E., Wood, J. M.:* Biochemistry (1971) (in press).
51. *Morningstar, J. F., Kisliuk, R. L.:* J. Gen. Microbiol. *39*, 43 (1965).
52. *Jaenicke, L.:* Biochem. J. *99*, 210 (1966).
53. *Cauthen, S. E., Foster, M. A., Woods, D. D.:* Biochem. J., *98*, 630 (1966).
54. *Brodie, J. D., Burke, G. T., Magnum, J. H.:* Biochemistry *9*, 4297 (1970).
55. *Burke, G. T., Magnum, J. H., Brodie, J. D.:* Biochemistry (in press) (1971).
56. *Brodie, J. D.:* Proc. Natl. Acad. Sci. U. S. *62*, 461 (1969).
57. *Weissbach, H., Peterkovsky, A., Redfield, B. G., Dickerman, H.:* J. Biol. Chem. *238*, 3318 (1963).
58. — *Redfield, B. G., Dickerman, H., Brot, N.:* J. Biol. Chem. *240*, 856 (1965).
59. *Guest, J. R., Friedman, S., Dilworth, M. J., Woods, D. D.:* Ann. N. Y. Acad. Sci. *112*, 774 (1964).
60. *Woods, D. D., Foster, M. A., Guest, J. R.:* Transmethylation and Methionine Biosynthesis, p. 138 (*S. K. Shapiro* and *F. Schlenk*, eds.). University of Chicago Press 1965.
61. *Guest, J. R., Helliener, C. W., Cross, M. J., Woods, D. D.:* Biochem. J. *76*, 396 (1960).
62. *Larrabee, A. R., Rosenthal, S., Cathon, R. E., Buchanan, J. M.:* J. Am. Chem. Soc. *83*, 4094 (1961).
63. — *Buchanan, J. M.:* Acta Chem. Scand. *17*, 288 (1963).
64. *Hatch, F. T., Larrabee, A. R., Cathon, R. E., Buchanan, J. M.:* J. Biol. Chem. *236*, 1095 (1961).
65. *Taylor, R. T., Weissbach, H.:* J. Biol. Chem. *242*, 1502 (1967).
66. — — Arch. Biochem. Biophys. *123*, 109 (1968).
67. — — Arch. Biochem. Biophys. *129*, 728 (1969).
68. — — Arch. Biochem. Biophys. *129*, 745 (1969).
69. *Rudijer, H., Jaenicke, L.:* European J. Biochem. *10*, 557 (1969).
70. *Wood, H. G.:* J. Biol. Chem. *194*, 905 (1952).
71. *Poston, J. M., Koratomi, K., Stadtman, E. R.:* Ann. N. Y. Acad. Sci. *112*, 804 (1964).
72. *Kuratomi, K., Poston, J. M., Stadtman, E. R.:* Federation Proc. *24*, 421 (1965).
73. — *Poston, J. M., Stadtman, E. R.:* Biochem. Biophys. Res. Commun. *23*, 691 (1966).
74. *Ljungdahl, L., Irion, E., Wood, H. G.:* Biochemistry *4*, 2771 (1965).
75. — — — Federation Proc. *25*, 1642 (1966).
76. *Lezius, A. G., Barker, H. A.:* Biochemistry *4*, 510 (1965).

77. *Blaylock, B. A., Stadtman, T. C.:* Ann. N. Y. Acad. Sci. *112*, 799 (1964).
78. *Wolin, M. J., Wolin, E. A., Wolfe, R. S.:* Biochem. Biophys. Res. Commun. *12*, 464 (1963).
79. *Wood, J. M., Wolfe, R. S.:* Biochemistry *5*, 3598 (1966).
80. *Blaylock, B. A.:* Arch. Biochem. Biophys. *124*, 314 (1968).
81. *Bernhauer, K., Friederich, W.:* Angew. Chem. *65, 627* (1953).
82. *Barker, H. A.:* Arch. Mikrobiol. *7*, 404 (1936).
83. *Bryant, M. P., Wolin, E. A., Wolin, M. J., Wolfe, R. S.:* Arch. Mikrobiol. *59*, 20 (1967).
84. *Wolin, E. A., Wolin, M. J., Wolfe, R. S.:* J. Biol. Chem. *238*, 2882 (1963).
85. *Wood, J. M., Allam, A. M., Brill, W. J., Wolfe, R. S.:* J. Biol. Chem. *240*, 4564 (1965).
86. — *Wolfe, R. S.:* Biochem. Biophys. Res. Commun. *19*, 306 (1965).
87. — *Wolin, M. J., Wolfe, R. S.:* Biochemistry *5*, 2381 (1966).
88. *McBride, B. C., Wolfe, R. S.:* Biochemistry *10*, 2317 (1971).
89. *Penley, M. W., DeSimone, R. E., Wood, J. M.:* Biochemistry (in press) (1971).
90. — *Brown, D. G., Wood, J. M.:* Biochemistry *9*, 4302 (1970).
91. *Roberton, A. M., Wolfe, R. S.:* Biochem. Biophys. Acta *192*, 420 (1969).
92. *Halpern, J., Maher, J. P.:* J. Am. Chem. Soc. *86*, 2311 (1964).
93. *Challenger, F.:* Chem. Rev. *36*, 315 (1945).
94. *McBride, B. C., Wolfe, R. S.:* Bacteriol. Proc. *130*, 1969 (1969).
95. *Lee, H. A., Abeles, R. H.:* J. Biol. Chem. *238*, 2367 (1963).
96. *Brownstein, A. M., Abeles, R. H.:* J. Biol. Chem. *236*, 1199 (1961).
97. *Frey, P. A., Karabatsos, G. L., Abeles, R. H.:* Biochim. Biophys. Res. Commun. *18*, 551 (1965).
98. *Zagalak, B., Frey, P. A., Karabatsos, G. L., Abeles, R. H.:* J. Biol. Chem. *241*, 3028 (1966).
99. *Babior, B. M.:* J. Biol. Chem. *245*, 6125 (1970).
100. — *Gould, D. C.:* Biochem. Biophys. Res. Commun. *34*, 441 (1969).
101. — *Li, T. K.:* Biochemistry *8*, 154 (1969).
102. *Bernhauer, K., Gaiser, P., Muller, O., Muller, E., Gunther, F.:* Biochem. Z. *333*, 560 (1961).
103. *Nowicki, L., Pawelkiewicz, J.:* Bull. Acad. Polan. Sci. Cl II *8*, 433 (1960).
104. *Johnson, A. W., Shaw, N.:* Proc. Chem. Soc. (London) 420 (1960).
105. — — Proc. Chem. Soc. (London) 447 (1961).
106. *Hogenkamp, H. P. C., Barker, H. A., Mason, H. S.:* Arch. Biochem. Biophys. *100*, 353 (1963).
107. *Yamada, R. H., Shimiyu, S., Fukui, S.:* Arch. Biochem. Biophys. *117*, 675 (1966).
108. *Hill, H. A. O., Pratt, J. M., Williams, R. J. P.:* Proc. Roy. Soc. (London) *288A*, 352 (1965).
109. *Firth, R. A., Hill, H. A. O., Pratt, J. M., Thorp, R. G.:* Chem. Commun. 1013 (1967).
110. *Cockle, S. A., Hill, H. A. O., Pratt, J. M., Williams, R. J. P.:* Biochem. Biophys. Acta *177*, 686 (1969).
111. *Bayston, J. H., Looney, F. D., Pilbrow, B. R., Winfield, M. E.:* Biochemistry *9*, 2164 (1970).
112. *Assoor, J. M., Kahn, W. K.:* J. Am. Chem. Soc. *87*, 207 (1965).
113. *Schrauzer, G. N., Lee, L. P.:* J. Am. Chem. Soc. *90*, 654 (1968).
114. *Bayston, J. H., King, N., Looney, F. D., Winfield, M. E.:* J. Am. Chem. Soc. *91*, 2775 (1969).
115. *Lee, L. P.:* Ph. D. Thesis, University of California, San Diego (1970).

116. *Bayston, J. H., Looney, F. D., Winfield, M. E.:* Australian J. Chem. *16,* 557 (1963).
117. *Hamilton, J. A., Blakely, R. L., Looney, F. D., Winfield, M. E.:* Biochem. Biophys. Acta. *177,* 374 (1969).
118. *— Tamada, R., Blakely, R. L., Hogenkamp, H. P. C., Looney, F. D., Winfield, M. E.:* Biochemistry *10,* 347 (1971).
119. *Buckman, T., Kennedy, F. S., Wood, J. M.:* Biochemistry *8,* 4437 (1969).
120. *Stone, T. J., Buckman, T., Nordio, P. L., McConnell, H. M.:* Proc. Natl. Acad. Sci. U. S. *54,* 1010 (1965).
121. *Firth, R. A., Hill, H. A. O., Pratt, J. M., Williams, R. J. P., Jackson, W. R.:* Biochemistry *6,* 2178 (1967).
122. *Pailes, W. H., Hogenkamp, H. P. C.:* Biochemistry *7,* 4160 (1968).
123. *Law, P. Y., Brown, D. G., Lien, E. L., Babior, B. M., Wood, J. M.:* Biochemistry *10,* 3428 (1971).
124. *Hoffman, B. M., Eames, T. R.:* J. Am. Chem. Soc. *91,* 5168 (1969).
125. *Hamilton, C. L., McConnell, H. M.:* Structural Chemistry and Molecular Biology (*Rich* and *Davidson,* eds). W. H. *Freeman,* 115 (1968).
126. *Kennedy, F. S.:* Ph. D. Thesis, University of Illinois (1970).
127. *Hill, H. A. O., Pratt, J. M., Williams, R. J. P.:* J. Chem. Soc. 2859 (1965).
128. *— Mann, B. E., Pratt, J. M., Williams, R. J. P.:* J. Chem. Soc. A564 (1968).
129. *Dolphin, D., Johnson, A. W.:* Chem. Commun. 495 (1965).
130. *Brodie, J. D., Poe, M.:* Biochemistry *10,* 914 (1971).
131. *Hayword, G. C., Hill, H. A. O., Pratt, J. M., Varston, N. J., Williams, R. J. P.:* J. Chem. Soc. 6485 (1965).
132. *Firth, R. A., Hill, H. A. O., Pratt, J. M., Thorp, R. G., Williams, R. J. P.:* Chem. Commun. 400 (1967).
133. *— — — Thorp, R. G., Williams, R. J. P.:* J. Chem. Soc. A2428 (1968).
134. *— — Mann, B. E., Pratt, J. M., Thorp, R. G., Williams, R. J. P.:* J. Chem. Soc. A2419 (1968).
135. *Hill, H. A. O., Morallee, K. G., Collis, R. E.:* Chem. Commun. 888 (1967).
136. *Randall, W. C., Alberty, R. A.:* Biochemistry *6,* 1520 (1967).
137. *Ludwick, L. M., Brown, T. L.:* J. Am. Chem. Soc. (in press) (1971).
138. *Pratt, J. M., Thorp, R. G.:* J. Chem. Soc. A187, (1966).
139. *Herlinger, A. W., Brown, T. L.:* J. Am. Chem. Soc. *93,* 1790 (1971).
140. *Pople, J. A., Schneider, W. Y., Bernstein, H. J.:* High Resolution NMR, Chapter 10. New York: McGraw-Hill 1959.
141. *Hill, J. A., Pratt, J. M., Williams, R. J. P.:* J. Theoret. Biol. *3,* 1707 (1962).
142. *Law, P. Y., Wood, J. M.:* Unpublished data.
143. *Wood, J. M., Kennedy, F. S., Wolfe, R. S.:* Biochemistry *7,* 1707 (1968).
144. *Saika, A., Slichter, C. P.:* J. Chem. Phys. *22,* 26 (1954).
145. *Doddrell, D., Allerhand, A.:* Proc. Natl. Acad. Sci. U.S. (in press) (1971).
146. *— —* Chem. Commun. (in press) (1971).

Received July 8, 1971

Molybdenum-Containing Enzymes

R. C. Bray

School of Molecular Sciences, University of Sussex, Brighton BN 1 9 QJ, Great Britain

J. C. Swann

Biochemistry Department, Monash University, Clayton, Victoria 3168, Australia

Table of Contents

I. Biological Importance of Molybdenum

Molybdenum is a trace element which, whilst present only at low levels, is important in the metabolism of many living species (1). It has been known for a quarter of a century to play an essential role in nitrogen fixation by micro-organisms (2), and during the last few years in particular, progress has been made in characterising the molybdenum-containing enzymes involved. A variety of higher plants also depend on the presence of molybdenum to utilize nitrogen *via* nitrate-reducing systems (1). In areas where there is too low a level of molybdenum in the soil, these plants show deficiency symptoms (3). In mammals, low molybdenum intake does not seem to have such clear-cut effects (4); nevertheless the metal has been shown to be an essential constituent of several mammalian enzymes.

Table 1. *Some molybdenum-containing enzymes. For some of the enzymes the numerical data are rather approximate only. Turnover numbers refer to temperatures between 23 and 30°. Since some of the molecular weights are particularly uncertain, contents of flavin etc. are expressed per molybdenum atom rather than per mole of protein. Most of the enzymes have a much wider range of substrate specificities than has been indicated*

		Typical substrates	
Name	Source	Reducing	Oxidizing
Nitrogenase[a]	*Clostridium pasteurianum*	Dithionite, ferredoxin	N_2
Nitrogenase[a]	*Klebsiella pneumoniae*	Dithionite	N_2
Nitrogenase[a]	*Azotobacter vinelandii*	Dithionite	N_2
NADH dehydrogenase	*Azotobacter vinelandii*	NADH[b]	Menadione
Nitrate reductase[c]	*Escherichia coli*	Reduced methyl viologen	NO_3^-
Nitrate reductase	*Neurospora crassa*	NADPH	NO_3^-
Nitrate reductase	*Aspergillus nidulans*	NADPH	NO_3^-
Sulphite oxidase	Bovine liver	SO_3^-	O_2
Aldehyde oxidase[f]	Rabbit liver	Aldehydes	O_2
Xanthine dehydrogenase	Chicken liver	Purines, aldehydes	NAD
Xanthine dehydrogenase	*Micrococcus lactilyticus*	Purines, aldehydes	Ferredoxin
Xanthine oxidase	Cow's milk	Purines, aldehydes	O_2

[a] Nitrogenase consists of two proteins, one containing iron and the other both iron and molybdenum. Data in the composition and molecular weight columns refer to the latter protein only.

[b] Also reacts more slowly with aldehydes.

II. Survey of Molybdenum-Containing Enzymes

Table 1 summarizes the properties of most of the known molybdenum-containing enzymes. A few general points may be noted. Molecular weights are all relatively high ($> 100,000$) and additional prosthetic groups, such as flavin, iron or cytochrome-b are invariably present. Molybdenum is firmly bound to the enzymes and apparently only in the case of nitrate reductase (21) has the metal been reversibly dissociated. The frequency with which a composition of two atoms of molybdenum per enzyme molecule occurs is noteworthy. This is reminescent of the pronounced tendency of low molecular weight molybdenum compounds to be bi-nuclear with two molybdenum atoms close together and interacting (e.g. 22). However this analogy may be purely coincidental, since there is in fact no positive evidence for Mo—Mo interaction in any of the enzymes.

Mo content in g. atom/ mole	Composition			Molecular weight	Turnover (mole/sec./ g. atom Mo)	Ref.
	Other constituents (Moles or g. atom/g. atom Mo)					
	Flavin	Iron	Cyt-b			
1	—	15	—	170,000	1	(6, 7a, 7b)
1	—	17	—	220,000	1	(8, 9)
2	—	36	—	270,000	1	(10)
?	FMN ($\frac{1}{2}$)	1	—	?	140	(11)
1	—	40	—	1,000,000	7,000	(12)
1–2	FAD[d]	?	present	230,000	50[e]	(13)
?	present	?	—	200,000	50[e]	(14)
2	—	—	b₅ (2)	110,000	?	(15)
2	FAD (1)	4	—	300,000	?	(16)
2	FAD (1)	4	—	300,000	8	(17)
2	FAD (1)	4	—	250,000	20	(18)
2	FAD (1)	4	—	275,000	12	(19, 20)

[c]) The particulate enzyme from which this solubilized form is derived contains cyt-b, and can utilize formate as reducing substrate.
[d]) Dissociates with K_m about 2×10^{-8} M.
[e]) Very approximate.
[f]) Also contains coenzyme Q.

Turnover numbers of molybdenum-containing enzymes generally tend to be low. A brief discussion of each of the enzymes in Table 1 is given below.

Nitrogenase is a complex enzyme which is currently attracting a great deal of interest (see refs *5a* and *5b* for recent reviews). Preparations from all sources, consist of two proteins, one containing iron and the other containing both iron and molybdenum. The exact relationship between these two subunits is not understood. The enzyme has proved exceptionally difficult to work with because of its sensitivity to oxygen, but is now available in considerable quantities and the molybdenum-containing fraction from one source has been crystallized (*10*). Hence one might hope soon, for substantial progress in this field.

The recently discovered molybdenum-containing NADH dehydrogenase from *Azotobacter vinelandii* (*11*) is produced instead of the more familiar iron-sulphur NADH dehydrogenase, only when the organism grows in iron-deficient media. Because it raises questions of what analogies might exist between molybdenum and iron-sulphur prosthetic groups, further work on this enzyme and a detailed comparison of it with the normal enzyme will be awaited with interest.

Nitrate reductases (*23*) have been prepared in small quantities and in varying states of purity from several micro-organisms, including *Neurospora crassa* (*13*), *Escherichia coli* (*12*), *Micrococcus denitrificans*[1]) (*24*), *Pseudomonas aeruginosa* (*25*) and *Aspergillus nidulans* (*14, 26*). There is evidence, in some cases indirect, that all contain molybdenum, presumably in their active centres but generally, little is known of its function in the catalytic reaction. On the other hand, there has been a great deal of interesting genetic work on nitrate reductases (*27, 28b, 29*) and though a detailed discussion is beyond the scope of this review, one conclusion from the work of *Nason* and co-workers must be mentioned. This is that nitrate reductase may have a subunit (possibly containing molybdenum, though this is not clearly stated by the authors) which is interchangeable with the corresponding subunit from other enzymes such as xanthine oxidase (*30*).

Of the mammalian enzymes, the sulphite oxidase of bovine liver has only recently been discovered to contain molybdenum (*15*). The better known molybdenum enzymes, xanthine oxidase from cows' milk (*31*) and aldehyde oxidase from rabbit liver (*16*) are closely related to one another as they are to the xanthine dehydrogenases from chicken liver (*17*) and from bacteria (*18*).

---

[1]) More recent work (*113*) casts doubt on the role of molybdenum in nitrate reductase from *M. denitrificans.*

III. Investigations of the Role of Molybdenum: Studies by Electron Paramagnetic Resonance

The precise function of molybdenum in the catalytic reactions of the various enzymes is of considerable interest both chemically and biochemically. In general, however, data bearing on this question are not readily obtainable, even for those enzymes which can be obtained in reasonable quantities at a high state of purity, because of the complexities of the enzymes and their low molybdenum contents. The metal is presumably bound to amino acid residues but the ligand atoms are unknown. Since molybdenum can exist in all oxidation states from 2 to 6, some oxidation-reduction function for it in the enzymes might be assumed. Also, since the metal forms a range of complexes (cf. *28a*), some role in substrate binding might be expected. One might further speculate that the relatively low turnover numbers found among molybdenum enzymes are related to slow ligand exchange reactions of the metal. However, in this respect there seem to be few relevant data on low molecular weight molybdenum complexes available for comparison (but see *32*).

Of experimental methods for studying the metal in enzymes, light absorption in the visible region from molybdenum chromophores is likely to be weak and frequently masked by stronger absorption from other enzyme constituents. Indeed only recently has a small molybdenum contribution to the absorption spectrum of even the most studied of these enzymes, xanthine oxidase, been demonstrated (*33*, see Section V F).

The physical method which has contributed most to studies on molybdenum-containing enzymes is electron paramagnetic resonance (EPR) (*34, 35*). Before considering results it may be useful to summarize the advantages and limitations of the method, when applied in this field. The great advantage is its specificity. Molybdenum EPR signals can generally be distinguished quite readily from those of other paramagnetic species present. Of the valencies of the metal, only Mo(V) and Mo(III) are paramagnetic and hence give signals. Although signals from these two valencies might not readily be distinguishable from one another, it seems likely that only those from Mo(V) have so far been seen in enzyme samples. Whilst appearance of an EPR signal provides evidence for the presence of Mo(V), absence of signals does not necessarily prove the absence of this valency, owing to the possibility of interaction of the metal with other paramagnetic species (cf. *36*). The sensitivity of the method is such that work at concentrations below about 10^{-5} to 10^{-6} M of a signal-giving species is very difficult. This means in practice that only enzymes available in rather large quantities (i.e. at least in tens of milligrams but

if possible in grams) are amenable to study by EPR. The usefulness of the
EPR method in enzyme studies has been enhanced by the development
of the rapid-freezing method (37, 38) which facilitates kinetic studies of
the pre-steady-state type. Thus apart from identifying Mo(V) in enzymes,
EPR can in principle give information about its concentration and about
the time course of changes in it during enzyme turnover. Additionally
it can give information on the symmetry of bonding around the metal
atom and about ligands and other nuclei which interact with it. Also if
molybdenum atoms are not all chemically equivalent to one another,
this can often be detected by EPR.

Studies bearing on the role of molybdenum in enzymes will be exemplified by a detailed summary of results on the most studied of these
enzymes, which is undoubtedly milk xanthine oxidase. To put this in
its context, it will be preceded by a review of the general properties of
xanthine oxidase. The final section will then be a short account of work
on some of the other molybdenum enzymes.

IV. Milk Xanthine Oxidase: General Properties

A. Introduction

Milk xanthine oxidase has a molecular weight of 275,000 (39) and contains 2 Mo atoms, 2 FAD and 8 Fe per mole. (See reference 31 for a
review of work on this enzyme up to 1962.) An unusually wide range of
oxidations is catalysed, reducing substrates including purines and related
heterocyclic compounds, as well as aldehydes, certain quaternary nitrogen
compounds such as N-methyl nicotinamide (40) and NADH. In all cases,
apart from with NADH, the reaction catalysed is a hydroxylation of the
substrate. (Aldehydes fall into this pattern, since they react in the non-
hydrated form (41).) The oxygen atom introduced into the substrate
molecule is derived from the medium, rather than from molecular oxygen
(42). The enzyme can readily be prepared in gram quantities in a high
state of purity (19) and in comparison with many enzymes is reasonably
stable. The above combination of properties has made xanthine oxidase
an attractive object of study to enzymologists throughout the present
century and work on it has been extensive. On the other hand the biological role of the enzyme is not clear[2] (31). Nevertheless the detailed
studies which have been performed on it should facilitate the under-

[2] Despite a previous suggestion (43) that xanthine oxidase has anti-tumour properties, re-evaluation (44) has shown that the original data were not statistically
significant.

standing of other related enzymes which may be of more immediate interest from a biological standpoint.

B. Non-Protein Components

The feature of xanthine oxidase which is no doubt of the greatest chemical interest, is the presence of several non-protein components. Much effort has been expended in attempting to elucidate the respective roles of iron, flavin and molybdenum in the various enzyme catalysed reactions. Numerous studies of the iron constituent have been made of late (45, 46, 47, 48, 49, 50), it having been found to be of the iron-sulphur (51 a, 51 b) type. Neither iron (19) nor molybdenum (31) can be removed reversibly from the enzyme, though the FAD can be (52, see below).

All three non-protein components show EPR signals during the catalytic cycle, and rapid-freezing EPR studies have shown that when the enzyme is treated with reducing substrates, the three components react with clearly differing time-courses (53, see also 54). This suggests strongly that the components cannot be in direct physical contact with one another on the enzyme surface. The exact mechanism whereby reducing equivalents are transferred intramolecularly from one component to another and the possibility that this transfer might be, in some way, through the protein then becomes of particular interest. On the rather similar enzyme, aldehyde oxidase, EPR saturation studies (55) have provided evidence for magnetic interactions between the constituents, suggesting that distances between them may be of the order of 10 to 20 Å. Analogies of the xanthine oxidase and aldehyde oxidase electron transfer systems to the respiratory chain make this a particularly significant area for future studies.

In the original rapid-freezing work on xanthine oxidase (53) it was found that in experiments employing about 1 mole of xanthine per mole of enzyme and an excess of oxygen, the time sequence of appearance of the various EPR signals was molybdenum (V), followed by flavin semiquinone radical (FADH), followed by iron. This suggested that the electron transfer sequence might be:

$$\text{xanthine} \rightarrow \text{Mo} \rightarrow \text{Flavin} \rightarrow \text{Fe} \rightarrow O_2$$

It must be realised however (as indeed it was at the time of these experiments (53)) that the use of such studies to determine the sequence, has certain limitations. The first limitation is that nothing can be said about the role of species not detectable by EPR, such as fully reduced flavin. Secondly and more generally, it has to be assumed for both the oxidizing and reducing substrates, that each reacts with only one of the electron transfer components in the enzyme and that there is a linear sequence of intramolecular redox reactions among these components. An additional complication is that the original work failed to distinguish between the FADH and O_2^- free radicals, both of which were presumably present (*56, 57*). It is thus hardly surprising that, as discussed below, some revision in the proposed sequence has since had to be made.

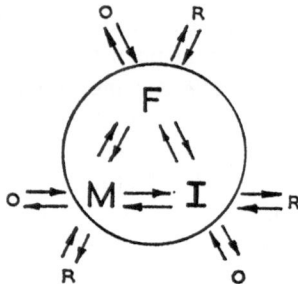

Scheme 1. Possible oxidation-reduction reactions between reducing and oxidizing substrate molecules (R and O respectively) and the molybdenum (M), flavin (F) and iron (I) of xanthine oxidase. The enzyme molecule is represented by the circle and arrows indicate transfer of reducing equivalents

   In principle, any or all of the reactions indicated in Scheme 1 might take place. In an extreme case, one could envisage a situation in which reducing equivalents entered and left the enzyme *via* one component only, leaving the other two components with no obvious function other than the storage of reducing equivalents. Indeed, this is apparently (*33, 115*) just the situation when xanthine is used as reducing substrate with phenazine methosulphate as oxidizing substrate, both of these reacting at the molybdenum site of the enzyme. More generally all the evidence (*52, 53, 59*) is in fact consistent (cf. section V F) with reducing substrates, other than NADH, reacting with the enzyme exclusively *via* molybdenum. Further, it has been shown within the last few years, that flavin may be dissociated reversibly from the enzyme, with concomitant loss of oxidase but not of dehydrogenase activity (*52*). This is good evidence that flavin is a binding site for oxygen. Nevertheless, it should perhaps be added, that xanthine oxidase reduces oxygen to both super-oxide (*57, 60*) and to hydrogen peroxide, the relative amount of these depending on the experimental conditions (*61, 62*). This suggested (*62*) that there may be more than one oxygen binding site in the enzyme and this possibility may not yet perhaps have been fully excluded, since oxidase activity of the deflavo enzyme seems to have been looked for only under a limited range of conditions (52).

The role of the iron-sulphur system of xanthine oxidase in the catalytic reaction is somewhat problematical. Nevertheless, it is clear, both from rapid freezing EPR (53) and from stopped-flow measurements monitored optically at 450 nm (58, 63) (where both iron and flavin are measured), that iron is reduced and oxidized at catalytically significant rates. Perhaps the best interpretation is that it functions as a store for reducing equivalents within the enzyme when this is acting as an oxidase, though it may well represent the main site of electron egress in dehydrogenase reactions (52).

C. Active and Inactive Forms

Since the literature on xanthine oxidase has been complicated for almost twenty years by controversy relating to the presence or absence of various inactive xanthine oxidase species as contaminants of the active enzyme, it may be useful to review this subject here. Such contamination was first proposed by *Morrell* (64), to explain rapid and slow phases in the anaerobic bleaching at 450 nm of the enzyme by the substrate xanthine. He found that the extent of the rapid phase only, was related to activity/A_{450} and hence to the proportion of active enzyme in the samples. *Avis et al.* (65, see also 66) showed that Activity/A_{450} and Activity/Mo varied independently of one another and proposed that there were two inactive species. Recent work (19, 20, 33) has clarified the situation considerably. The two inactive species which have to be considered (19) (though existence of further species is not entirely excluded) are termed *demolybdo xanthine oxidase* and *inactivated xanthine oxidase*[3]). The latter, only, is a universal contaminant of xanthine oxidase samples. De-molybdo xanthine oxidase is a natural constituent of milk, apparently produced in increased quantities by cows on a relatively poor nutritional regimen, and may be removed from preparations of the enzyme by a selective denaturation procedure utilizing high concentrations of sodium salicylate (19). (The decreased stability of this species possibly suggests that molybdenum plays some part in maintaining the native structure of the enzyme as well as participating in its catalytic function.) The demolybdo form was absent from the preparations reported by

---

[3]) In earlier work (66) where the presence of two inactive species was recognised but their origins less fully understood, these species were termed xanthine oxidase-i_1 (now inactivated xanthine oxidase) and xanthine oxidase-i_2 (now de-molybdo xanthine oxidase). The former species is referred to by *Massey* and co-workers as xanthine oxidase with 'non-functional active sites' (33).

Massey and co-workers (*58*), presumably owing to differences in starting material. Inactivated xanthine oxidase on the other hand, is a preparation or storage artefact (*19, 33*), whose presence may be minimized by rapid manipulation of the enzyme, by working in the presence of EDTA (*68*) and by storage at liquid nitrogen temperatures (*67*). Though the most active xanthine oxidase samples prepared to date (*69*) may contain as much as 90% of active enzyme, contamination by other forms has in the past generally been much higher, this amounting apparently to about 30% on samples of *Massey* and co-workers (*58*) and to 50—60% on those of earlier workers (*66, 70, 71*).

The question then arises of whether, and if so to what extent, such universal contamination has affected the overall properties of preparations of xanthine oxidase. The answer appears (*67*) to be essentially the one given by *Morell* (*64*), namely that at short reaction times the inactive forms are inert. Thus, up to reaction times of seconds or more, the presence of varying quantities of inactivated xanthine oxidase has no detectable influence on the course of reduction of the active enzyme by substrates. Under these conditions, then, inactive forms serve merely as a diluent of the active enzyme and, providing the concentration of the latter can be determined properly, their presence should have no adverse consequences in studies on the catalytic mechanisms. Stopped-flow and rapid freezing EPR are therefore valid methods for such work. (It is noteworthy that the latter method was originally developed (*37, 38*) specifically for work on xanthine oxidase, in anticipation of the above conclusion.) On the other hand, later work (*67*) has confirmed fully that inactivated xanthine oxidase plays a real and complex part in the slower reactions which go on in the presence of the substrate. These reactions, involving molybdenum, are discussed later (Section V D). An important conclusion is that the difference between inactivated xanthine oxidase and the active enzyme may represent no more than a change in one of the ligand atoms of the molybdenum (see also Section V B).

A failure by one of us to take fully into account the presence of inactivated xanthine oxidase, leading to misinterpretation of incomplete reaction of enzyme with iodoacetamide and hence to the apparently erroneous conclusion, that the two FAD molecules in the enzyme were non-equivalent (*72*), may serve as a warning to others. This reagent has since been shown to alkylate the flavin of reduced xanthine oxidase molecules, whether these are of the active or inactivated forms (*73*). Thus, under conditions where little of the inactivated form is reduced, the reagent becomes a specific one for the active enzyme (*20*). In the original experiments (*59, 72*) the content of active enzyme was, by coincidence, rather close to half of the total enzyme present. Thus, the presence of inactivated enzyme, rather than a lack of reactivity of one of the two flavins of the active enzyme molecule, was the cause of the observed results. Clearly, one should be on the look out for the presence of inactive variants in all enzyme chemistry studies, perhaps particularly so when working with molybdenum-containing enzymes.

A further controversy in the literature on xanthine oxidase preparations may be mentioned here. This concerns use of proteolytic enzymes in the purification procedure, a step first introduced by *Ball* (74), which undoubtedly increases yields of purified enzyme and generally simplifies the preparation. Despite suggestions to the contrary (70) it now seems (19, 58) that this treatment has little or no effect on the properties of the purified enzyme.

V. Milk Xanthine Oxidase: Studies on the Role of Molybdenum

A. Types of Molybdenum(V) EPR Signal from the Reduced Enzyme

Since studies bearing on the role of molybdenum in milk xanthine oxidase have depended heavily on the EPR method, it is convenient to precede detailed discussion by a general description of the various molybdenum(V) EPR signals which have been obtained from the enzyme.

'Resting' oxidized milk xanthine oxidase gives no EPR signals (53, 75), but under specific reducing conditions a considerable variety of Mo(V) signals appear and some of these are illustrated in Fig. 1. However, on strong reduction of the enzyme (e.g. with excess sodium dithionite for some hours), molybdenum signals disappear virtually completely (45, 49, 76, 77). In general, spectra obtained from the enzyme are due to signals from molybdenum atoms in a number of different chemical environments, superimposed on one another. The resultant spectra are therefore complicated. By appropriate choice of reducing conditions it has however been possible to obtain most of the individual signals on their own. Spectra in Fig. 1, are (with the exceptions noted in the caption) each believed due to an individual species.

Nomenclature of the signals has presented some difficulties. Originally (71) prominent peaks in the spectrum were given the letters α, β, γ, δ, as shown in Fig. 1. However, as more types of signal were discovered, this system became unsatisfactory because of signal overlap and particularly because several different signals all have prominent peaks in the γ–δ region. *Bray* and *Vänngård* (78) then attempted to use the time required for signal development as a basis for naming the signals. Though their system has certain obvious disadvantages, it will be adhered to here, if only to avoid further complicating the literature by introducing a new system. The time necessary for development of the various signals, under the specific reducing conditions which have been employed, differs greatly from one signal to another, ranging from a few milliseconds up to hours or even days. Accordingly, signals are referred to as the *Very*

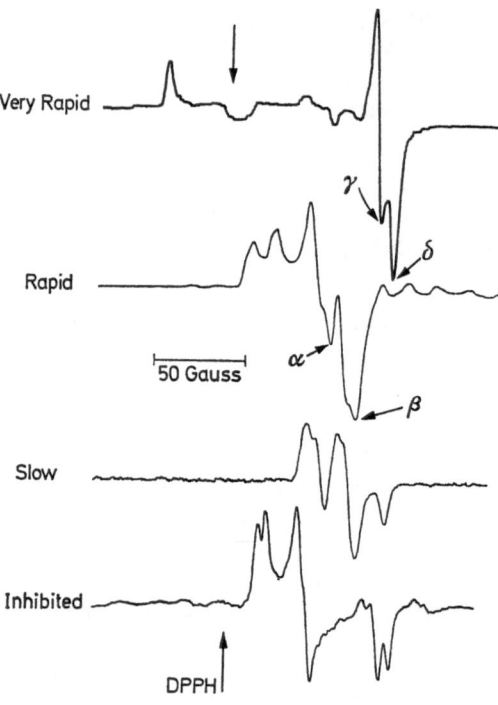

Fig. 1. Types of Mo(V) EPR spectra from reduced milk xanthine oxidase. All were recorded at about −150° on a 9.3 GHz instrument and are shown with magnetic field increasing from left to right. The reducing agents and conditions of reduction (generally at about 20°) were as follows:

Very Rapid: xanthine, 20 m.sec. at pH 10,
Rapid: dithionite, 1 sec. at pH 8.2,
Slow: dithionite, 30 min. at pH 8.2,
Inhibited: dialysed against aerated 1 M methanol containing salicylaldehyde for several days at 5°.

Apart from small features in the centre of the Very Rapid signal, which are due to slight contamination with the Rapid signal, all spectra consist of only one signal type. The Very Rapid, Slow and Inhibited signals, as illustrated, are all believed to represent single chemical species. The Rapid signal is believed due to two species with slightly different parameters, present in approximately equal amounts (see text). (Modified from ref. *59*)

Rapid, Rapid, Slow and *Inhibited* signals, respectively, Parameters of these signals are summarized in Table 2. It should be noted particularly that the Very Rapid and the Rapid signals develop within times comparable to the turnover time of the enzyme. They would therefore be expected to represent intermediates in the catalytic reaction. On the other hand, the Slow and the Inhibited signals take very much longer to appear and can have no such direct relationship to catalysis. It might still, however, be hoped that study of these two signals would lead to useful information relating to active site structure.

An important feature of three of the four signal types, is interaction of molybdenum with protons (*78, 79, 80*). This gives rise to a characteristic doublet structure in the spectra. The doublets in the Inhibited signal are particularly obvious (Fig. 1). Proton splittings of up to 16 gauss are observed (Table 2). No analogous molybdenum-proton interaction seems to have been reported in low molecular weight molybdenum compounds, though metal-proton interaction is well known in other fields, for example, Cu(II)-proton interaction in complexes of the salicylaldimine type (*82*). Origins and possible locations of the various protons interacting with molybdenum will be discussed in detail below. No hyperfine interactions with nitrogen, a nucleus which can cause splitting into triplets, have been observed in the spectra.

Conditions under which the various signals are obtained are summarized in Scheme 2, together with data on the enzyme species involved, to be discussed below. The Very Rapid signal is obtained on reaction with xanthine, particularly at high pH values, at reaction times in the range of about 5—100 m.sec. only (for 20—25 °) (*53*). The only other substrate which definitely yields this signal is 6—methylpurine (*54*). However, here reaction times of minutes are required at pH 10 (*83*). The Rapid signal is obtainable with all substrates tested to date. In some cases it develops fully in as little as 50 m.sec. (*54*) and under anaerobic conditions it is stable for hours or even days (*67*). The Slow signal is obtained on treating enzyme with moderate amounts of dithionite, for times in the region of 20 min. (*76*). It is also obtainable at longer times, when reduction is carried out with substrates such as purine (*67, 76*), salicylaldehyde (*76*) or xanthine (*67*). The Inhibited signal develops during a specific reaction (in which xanthine oxidase loses its catalytic activity), brought about either by methanol (*54*) or by formaldehyde (*81*). It is worth noting that all signal-giving forms of xanthine oxidase arise on treatment of the enzyme with reducing agents, never under oxidising conditions. This constitutes possibly the strongest evidence that molybdenum is in its highest oxidation state, Mo(VI), in the resting enzyme, and that all the reduced, paramagnetic species correspond to Mo(V).

119

Table 2. EPR parameters of Mo(V) species from milk xanthine oxidase. Parameters have been measured from spectra of frozen solutions at about −150°, in some cases at both 9 and 35 GHz. Correctness of some of the interpretations was confirmed by computer simulations

Signal		g-values				¹H splitting (gauss)			⁹⁵Mo splitting (gauss)			Ref.
		1	2	3	Av.	1	2	3	1	2	3	
Very Rapid		2.025	1.956	1.951	1.977	—	—	—	41	24	37	(79)
Rapid, No Complex	A	1.991	1.968	1.963	1.974	14, 3(?)	14	14				(78)
	B	1.991	1.966	1.963	1.973	14, 3(?)	14	14				(78)
Rapid, Complex	I	1.989	1.969	1.964	1.974	12, 4	12, 0	12, 0	64			(80, 78)
	II	1.994	1.968	1.961	1.973	12, 12	12, 12	12, 12				(78)
Slow		1.975	1.970	1.957	1.967	16	16	16			70	(80)
Inhibited		1.989	1.977	1.953	1.973	4.4	3.9	5.6	57(?)	25	57(?)	(81)

B. Inactivated Xanthine Oxidase and the Slow Signal

The Slow signal was first reported some five years ago, but the question of its origin has caused considerable confusion (*54, 76,* cf. *49*). Recently, following a resurgence of interest (*19, 33, 73*) in inactivated xanthine oxidase, clear evidence has been presented (*20*) that the Slow signal is due, not to active xanthine oxidase as had been assumed, but to the inactivated form of the enzyme (cf. Scheme 2). It differs in this respect from the other types of signal. Thus, whereas intensity of the Rapid signal, generated under suitably controlled conditions, increases with increasing Activity/A_{450}, i.e. with increasing amounts of active enzyme, intensity of the Slow signal decreases (Fig. 2). This relationship was not observed in earlier studies because of the dual complications of having two inactive forms of the enzyme present in the samples originally tested, plus a failure to appreciate effects (*76*) of dithionite concentration and of reaction time on Slow signal intensity.

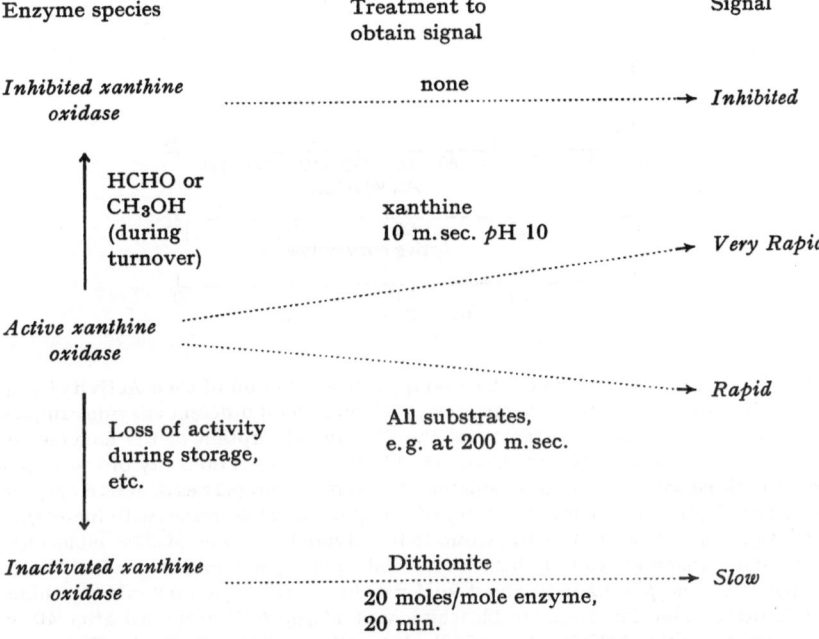

Scheme 2. Species responsible for Mo(V) signals from xanthine oxidase. The diagram illustrates typical treatments which may be used to obtain the various molybdenum EPR signals from milk xanthine oxidase. Nomenclature of the signals is that of reference (*78*) and of the enzyme species, that of reference (*19*). Conditions for signal development refer to *p*H 8.2 and 20—25 ° with about 0.1 mM enzyme unless otherwise stated

It is of considerable interest, in attempting to understand catalysis by xanthine oxidase, to compare the properties of the active enzyme with those of the inactivated form. Radioactive tracer and other techniques have provided evidence (*33*) that (at least in its reduced form) the

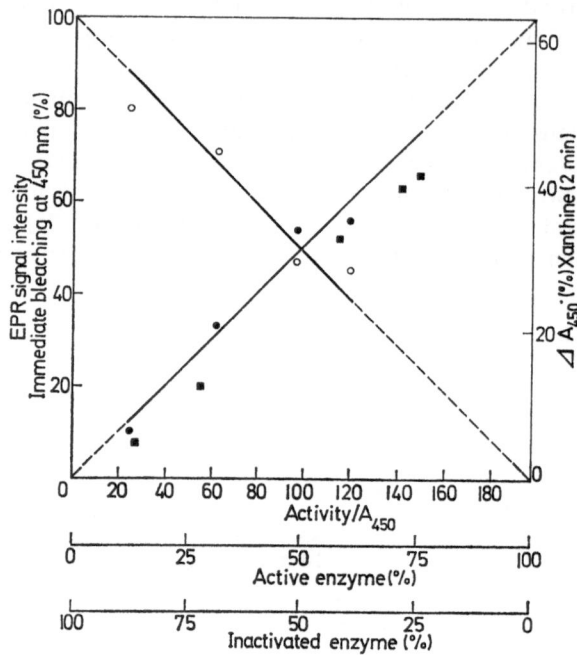

Fig. 2. Properties of xanthine oxidase samples as a function of their Activity/A_{450} ratios. Measurements of three parameters, each on several different enzyme samples are plotted. 'Immediate' bleaching of the 450 nm absorption of the enzyme by xanthine increases with Activity/A_{450} (cf. *64*). So, too, does intensity of the Rapid Mo(V) EPR signal. These two parameters are therefore properties of active enzyme molecules. On the other hand, intensity of the Slow signal decreases with increasing Activity/A_{450}, indicating this to be due to inactivated xanthine oxidase molecules. 'Immediate' anaerobic 450 nm bleaching (solid squares) was measured after 2 min. at about 22 ° and *p*H 8.2 with about 7μM xanthine oxidase and 0.2 mM xanthine. The limiting value for complete bleaching was ΔA_{450} : 63% (attained after 40 hr by a sample with Activity/A_{450} : 149). Intensities of the EPR signals are in arbitrary units. The Rapid signal (solid circles) was measured after 1.5 min. reduction at *p*H 8.2 and 20 ° by 2.0 mole xanthine and the Slow signal (open circles) after 20 min. at the same *p*H and temperature with 25 mole dithionite. The limiting Activity/A_{450} value indicated on the graph for pure active enzyme is 197 at 23.5 °. (This corresponds to the highest Activity/Mo observed by *Hart* and *Bray* ref. *68*.) (Data replotted from ref. *20*)

inactivated enzyme lacks a binding site for the inhibitor, alloxanthine[4]) and hence that it presumably lacks the binding site for reducing substrates also. A further difference may be noted in rates at which molybdenum of the active and inactivated enzyme, respectively, are reduced to Mo(V) by dithionite and re-oxidized by oxygen. For the active enzyme, such reduction (to give the Rapid signal) has $t_{\frac{1}{2}}$ about 1 sec. (85) and reoxidation $t_{\frac{1}{2}}$ 20 m. sec. (53), whereas for the inactivated enzyme, giving the Slow signal, the corresponding times are about 5 min. and 1 sec. (76). (These data refer to pH 8.2 and 20 ° with concentrations of the order of 0.1 mM enzyme and 1 mM dithionite or oxygen.) On the other hand, it seems that NADH, which reduces the enzyme via flavin (52), behaves in much the same way towards either active or inactivated enzyme (67). Further the EPR parameters of the Rapid and Slow signals, are not very greatly different from one another (Table 2), the main difference being a change from $g_{\parallel} > g_{\perp}$ to $g_{\perp} > g_{\parallel}$, with a small decrease in g_{av}.

The overall implications are that the differences between active and inactivated xanthine oxidase are not great. Inactivated xanthine oxidase seems to differ from active xanthine oxidase only in that molybdenum in the active centres has become nonfunctional and quite possibly this is brought about by the metal atoms undergoing no more than a change in a single ligand. Combining the above data with recent work by *Massey* and *Edmondson* (86) on cyanide inactivation (see Section V G) has led to the speculation (67) that the sulphur atom lost in cyanide treatment might perhaps be a ligand of molybdenum.

C. Multiple Forms of the Rapid Signal and their Origins

As noted in Section V A, every substrate which has been tested to date, has, under suitable conditions, given the Rapid signal. However, as indicated in Table 2, the Rapid signal is not a unique signal but a whole family of closely related signals. Detailed examination of these has provided interesting information bearing on the catalytic reactions of the enzyme. EPR spectra, obtained at 35 GHz as well as at 9 GHz, were analysed with the aid of computer simulation techniques (78). Since the protons of the Rapid signals exchange with those of the solvent water, both hydrogen and deuterium forms of the spectra could readily be examined. Rapid signals from xanthine, purine and salicylaldehyde could be accounted for satisfactorily, only, by assuming that there were at

---

4) Very recently (84) this specific binding of alloxanthine-like compounds to active xanthine oxidase has been made the basis of a method of separating the active from the inactivated enzyme, using a pyrazolo-(3,4) pyrimidine attached to agarose.

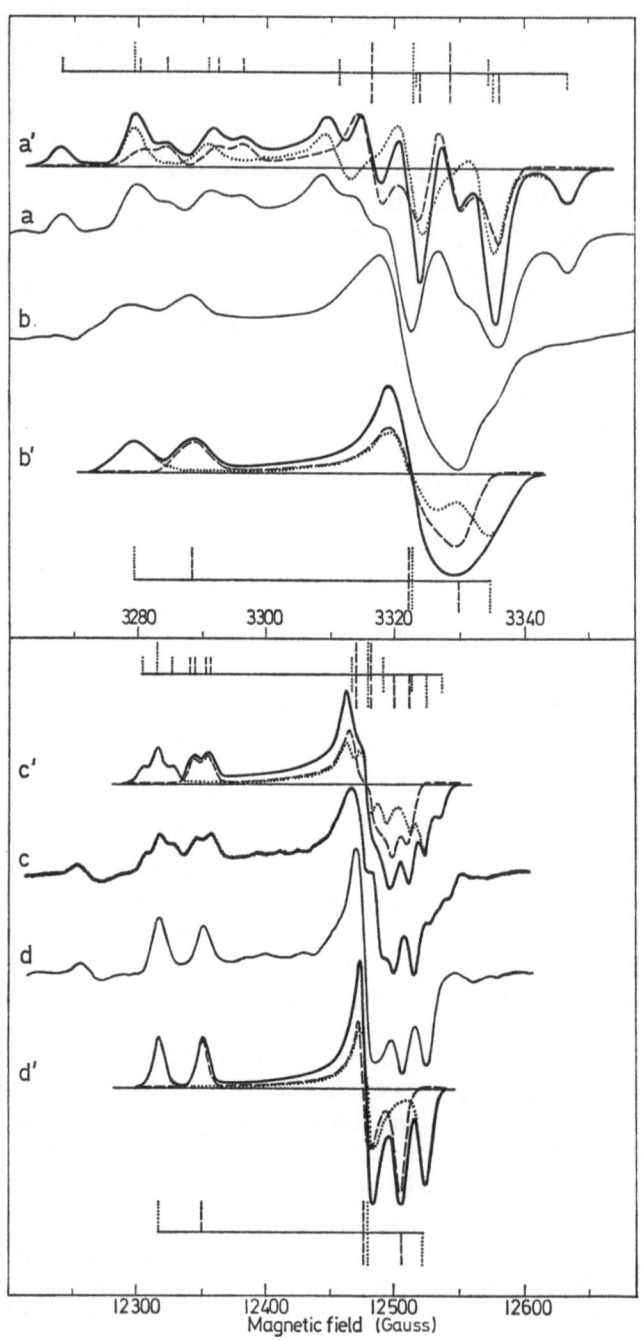

least two closely related but distinguishable signal-giving species present in all cases. Such analysis of Rapid signals is illustrated in Fig. 3, for spectra obtained with xanthine.

Before details of the xanthine Rapid signals are discussed, it is convenient to consider those from a group of substrates which includes salicylaldehyde, formaldehyde, and dithionite, all of which give Rapid signals indistinguishable from one another (76). Analysis (78) (for the salicylaldehyde spectrum) indicated two components, present in approximately equal amounts. It was considered that dithionite and its oxidation products, were unlikely to form complexes with the enzyme. The two spectral types must both therefore have arisen from uncomplexed, reduced, active enzyme molecules. They were designated (78): 'Rapid signal, no complex detected, type A and type B' (Table 2). It was suggested that these two signals corresponded to the two molybdenum atoms of the xanthine oxidase molecule. As will be discussed below, the proposal (87) that the active site of the enzyme contains two molybdenum atoms is now considered less likely than that there are two, essentially independent active centres, each having a single molybdenum atom. Thus, the presence of two signals implies that the chemical environment of the two centres in the reduced enzyme may not be quite identical.

A different type of phenomenon is observed for the two specific Rapid signals which are obtained from xanthine (Fig. 3). Whilst the 'no complex detected' types of signal are still seen with very low xanthine concentrations (88), these are replaced at higher xanthine concentration by a different signal (76, 88, see also Section V D and Fig. 4) which seems to represent a mixture of variable amounts of two complexes of reduced enzyme (76, 78). These complexes are, apparently, not with a product derived from the xanthine molecule which originally reduced molybdenum (76). Instead they involve a further xanthine molecule, this remaining un-oxidized in the complex. Ultimately, when molybdenum has been reoxidized (via iron and flavin), this substrate molecule having

---

Fig. 3. Rapid Mo(V) EPR signals obtained on reducing xanthine oxidase at pH 10 with 15 moles of xanthine for 1 min. at about 20°. The upper four spectra are at 9.1 GHz and the lower four at 34.4 GHz. a, a', c, c' refer to H_2O as solvent and b, b', d, d' to D_2O. a', b', c', d' are computer simulations of the experimental spectra, a, b, c, d, respectively. The interpretation is that two species, each having exchangeable protons which interact with Mo(V), are responsible for the signals. For one of these (dotted: complex type II) there are two equivalent interacting protons and for the other (dashed: complex type I), two non-equivalent protons. These species are believed to correspond to two different complexes of reduced xanthine oxidase with xanthine. (Reproduced from ref. 78; see also Table 2 for the parameters of the signals.)

remained at the active site, is itself then presumably ready to undergo oxidation. In other words, the complex might be thought of as a 'pre-Michaelis' state. The precise difference between the two xanthine complexes (whose signals are (78) referred to as the 'Rapid signal, complex detected, type I and type II': Table 2), is at present not clear. Other purines form complexes with the reduced enzyme, though the EPR spectra obtained are less characteristic than are those from xanthine. Complexes have been observed with purine itself (78) and with 1-methylxanthine (89), the latter apparently forming two complexes whose spectra are analogous to, but not identical with, those from xanthine.

D. Quantitative Studies on the Rapid Signal and Molybdenum Valency Changes

When xanthine oxidase is treated anaerobically with xanthine at pH 8.2, the signals obtained depend in a complex way both on enzyme and substrate concentrations and on reaction time. Detailed examination of

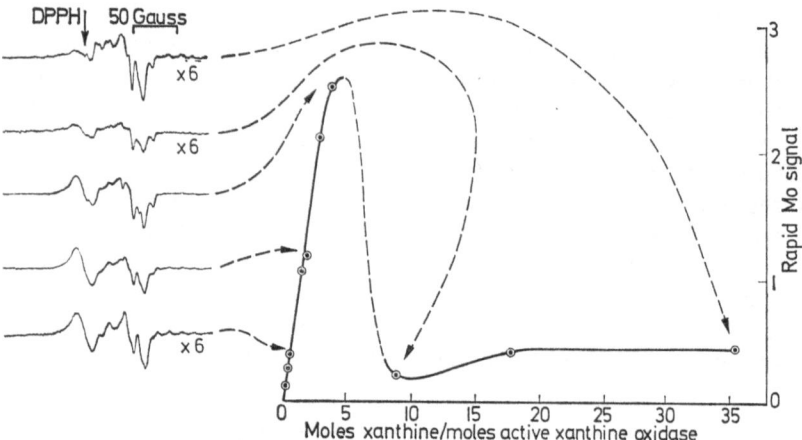

Fig. 4. Anaerobic titration of xanthine oxidase with xanthine at pH 8.2 with a reaction time of 2 min. at about 20°. The integrated intensity of the Rapid molybdenum EPR signals (in arbitrary units) is plotted against the number of moles of xanthine added per mole of *active* enzyme. Activity/A_{450} for the enzyme samples used was 112 corresponding to an active enzyme content of 57%. Thus the molar ratios of xanthine/total xanthine oxidase have been multiplied by 1.76 to refer to the active form only. Some of the EPR spectra (recorded at about $-130°$ and 9.3 GHz) are reproduced to show the changes in signal type as the amount of xanthine is increased. (Data re-calculated from ref. 88, with intensities corrected for variations in tube diameter and enzyme concentration calculated in terms of active enzyme.)

these processes has helped to clarify the question of which valency states of molybdenum are involved in catalysis. The available data will be summarized.

Anaerobic titrations of the enzyme with xanthine (*49, 53, 66, 88*) have all shown behaviour of the type illustrated in Fig. 4. At reaction times of the order of 1 or 2 minutes, only the Rapid signal is seen. As the amount of xanthine is increased, the intensity of this signal at first increases, then falls off almost to zero. Finally, a further slight increase occurs at high substrate levels. In Fig. 4, maximal intensity is achieved for about 4 moles xanthine per mole *active* enzyme, i. e. 2 xanthine per molybdenum. As indicated, the form of the Rapid signal changes during such titrations. No doubt (as discussed in the previous section) this is due to changes in the extent and nature of complex formation between the reduced enzyme and xanthine molecules[5]).

In experiments at a fixed xanthine concentration, the time-course for signal changes is correspondingly complicated. With an excess of xanthine, a whole series of changes takes place, starting at a reaction time of a few millisec. (*53*) and being scarcely complete even after two days (*67*). As noted above, the first signal to appear, as a transient, is the Very Rapid signal. The significance of this will be considered later. A typical time-course for intensity changes in the Rapid signal is illustrated in Fig. 5. This signal develops within about 200 m.sec. (Phase I), then 'fades' to around half its original intensity within about 1 sec. (Phase II). Subsequently, it increases once again (Phase III), to levels considerably higher than those attained in Phase I, this process requiring up to 2 hours for completion. Finally, at still longer times (Phase IV), the Slow signal from inactivated enzyme develops also, with some accompanying decrease in Rapid signal intensity. Of these phases, only Phase I takes place in a time which is comparable to the turnover time and is therefore of direct significance in catalysis. The rate of Phase II is highly concentration-dependent, depending, though, only on the concentration of active enzyme (*67*). On the other hand, the rate of Phase III depends markedly on the amount of inactivated enzyme present, reaction becoming progressively slower as Activity/A_{450} of the samples increases (*67*). Intensities of the molybdenum signals in processes such as those described above and indeed under any conditions, never exceed about half of the maximum intensity which would be expected if all the molybdenum in the samples was contributing to the signal (*67, 77*).

---

[5]) Under the conditions of these titrations Fe/S and flavin are partly reduced as well as Mo (*53*). At the maximum in Fig. 4, most of the xanthine is used in reducing the enzyme but calculations confirm that some should remain unreduced and available to form complexes.

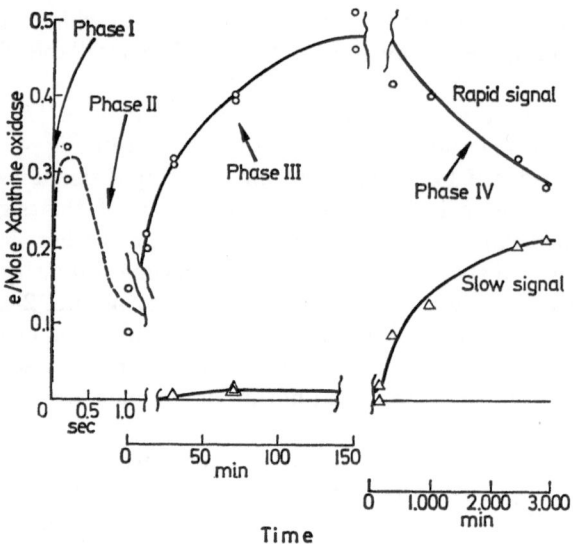

Fig. 5. Multiple phases in the reduction of xanthine oxidase by xanthine at pH 8.2. Intensities of the Rapid (circles) and Slow (triangles) molybdenum EPR signals expressed as electron/mole enzyme (i.e. per 2 atom Mo) are plotted as a function of time. Note the changes in the time scale. Rapid freezing was used for reaction times (at 22 °) up to 1 sec. and manual mixing for longer times (at 25 °); enzyme concentrations (immediately after mixing) were 0.09 mM and 0.13 mM respectively. The enzyme had Activity/A_{450}: 125 corresponding to 63% of active enzyme and 20 mole xanthine/mole enzyme was used. (Data from ref. *67*.)

Various explanations of the above phenomena have been considered (*59*) but it now seems (*67*) that the simplest is the best. All results seem consistent with molybdenum being Mo(VI) in the oxidized enzyme, which is reduced to Mo(V) and Mo(IV) by substrates. Incomplete Mo(V) signal development then depends at least in part on equilibria among the three valence states. Particularly, it seems clear that the drop in signal intensity in the titration experiments (Fig. 4), in the region of 7 moles xanthine/mole active xanthine oxidase and the Phase II 'fading' (Fig. 5) correspond to reduction to Mo(IV). The reality of this valence state in xanthine oxidase has been shown (*33*) by titration experiments on the complex which reduced enzyme forms with alloxanthine (see Section V F). Alternative explanations (*59*) for the intensity changes would involve spin-spin interaction of Mo(V), either with the second molybdenum of the enzyme or with other paramagnetic centres in it. Though such interactions may indeed exist in xanthine oxidase under some conditions, it seems that they cannot be important in relation to the present phenomena. From the theoretical work by *Leigh* (*36*) it seems that changes in spin-spin inter-actions should, in the case of a spin of low symmetry such as that of the molybdenum in xanthine oxidase, result in changes in form as well as in intensity of signals. In fact only very slight, if any, change of form of the Rapid signals is observed in critical regions such as between 4 and 9 moles xanthine in the titration experiment (Fig. 4), and none during Phase III (*67*). In further support of Phase II 'fading' being due to reduction to Mo(IV), rapid freezing EPR experiments (*67*) with allopurinol may be cited. Rapid signals appear in the normal way with this inhibitory sub-strate but soon disappear completely, i.e., Phase II goes to completion in this case. This is clearly to be equated with inhibition related to reduc-tion to Mo(IV) which then forms complexes with alloxanthine, the oxida-tion product from allopurinol (*33*, see Section V F).

As has been mentioned above, integrated signal intensities for molyb-denum have always been less than 1 g. atom of Mo(V) per mole of xanthine oxidase. However, there are indications from recently performed inte-grations (*90*) that observed intensities of the Inhibited signal, in which Mo(V) is known to be stabilized (*81*), can be accounted for quantitatively when due allowance is made for the other species present in the samples. If this is confirmed it should make possible final rejection of earlier suggestions (*87*) that the enzyme contains two interacting molybdenum atoms in a single active centre. It should also help to eliminate possibilities (cf. *78*) that only one of the two molybdenum atoms of the molecule is ever detected by EPR spectroscopy.

We have not so far mentioned the Phase III increase in the Rapid signal (Fig. 5). It seems (*67*) that Phase II represents 'over reduction' of molybdenum to Mo(IV), possibly by substrate radicals (see Section V H). The system then comes towards thermodynamic equilibrium by inter-action between reduced active enzyme molecules and oxidized inactive ones (*67*, cf. *64*). As Mo(IV) of the former is oxidized to Mo(V), during Phase III, so iron or flavin of the inactive enzyme is reduced. Later, in Phase IV, molybdenum of the inactive enzyme is reduced also to give the Slow signal. Alloxanthine, which as noted above, forms a stable complex with Mo(IV), seems to abolish both the slow phase in the 450 nm bleaching of the enzyme by xanthine and the Phase III increase in Rapid signal (*91*).

E. Origins of the Interacting Protons

As noted above (Table 2), the Rapid, Slow and Inhibited signals all show interaction of Mo(V) with one or more protons. Knowledge of the origin and location of these interacting protons is potentially helpful in under-standing the catalytic mechanism of the enzyme.

Protons in the Rapid and Slow signal-giving forms are exchangeable with those of water. Thus, in general, if these signals are developed in D_2O rather than in H_2O, proton doublets are replaced by single deuterium lines (the latter representing unresolved triplets, since hyperfine splitting by deuterons is much smaller than by protons) (78, 80, see also Fig. 3). Rapid signals developed at short reaction times will be considered later.

The Inhibited signal shows a single interacting proton and this, on the other hand, is non-exchanging. As was noted earlier, this signal is obtained (87) when enzymic activity is destroyed by specific treatment with methanol or formaldehyde. The Inhibited signal differs also from the other signals, in that it is stable in the presence of air. It may, however, be eliminated and enzyme activity restored by, e.g., treatment with dithionite (87). By employing deuteromethanol (80) or deuteroformaldehyde (87) to produce the signal, the proton has been shown to be a carbon-bound one from the inactivating agent. Presumably then, in the species giving the Inhibited signal, the interacting proton remains bound to the carbon atom of the reagent from which it was derived. It has been tentatively suggested (87) that the reaction is one of formylation of the enzyme. The precise location, however, of the formyl or other group in the active centre remains unknown.

From anisotropy of the proton splittings in the various signals, it should be possible to deduce information about molybdenum—proton distances. Since anisotropy in the main splitting of the Rapid signal is small, by assuming all splittings to be of the same sign, it was concluded (78) that the distance in this case must be greater than about 3 Å. Thus molybdenum hydrides can apparently be excluded from consideration. A comparable distance is indicated in the case of the Inhibited signal (87). Actual locations of some of the protons will be discussed below.

It has been found that the exchange of protons seen in the Rapid signal with those of the solvent, though this occurs in considerably less than the enzyme turnover time, is nevertheless sufficiently slow to be followed by the rapid freezing EPR method. Thus, the interesting observation has been made (59, 83, 87, 92), that if one employs substrate molecules deuterated specifically on the carbon atom at which oxidation occurs, then the Rapid signal which appears, comes up initially in its deuterium, rather than in its hydrogen form. The phenomenon has been observed for xanthine-8-^{2}H (83, 92) for 1-methylxanthine-8-^{2}H (59, 83, 87), less definitely for salicylaldehyde ($HOC_6H_4C^2HO$) (59, 87) and also in the reverse direction, for non-deuterated 1-methylxanthine in D_2O (83, 87). At longer reaction times, e.g. above about 1 sec. at 10 °C, only the hydrogen form of the Rapid signals is present when the solvent is H_2O. It was concluded (92) that the phenomenon represents direct transfer of a hydrogen atom from the substrates to the enzyme, followed by its exchange with solvent protons.

It should be noted that even at the shortest reaction times, the pure deuterium form of the signal is never observed for a deuterated substrate because the exchange process takes place at a comparable rate to that of signal development. Furthermore, 'warming-up' experiments (cf. 76) have demonstrated (83) that deuterium-hydrogen exchange can take place in the frozen state, down to about $-80°$. This implies that the rapid freezing technique might not be fully effective in quenching such a reaction. It was suggested earlier (87) that the observation of about 50% deuterium in the signals under such conditions provided evidence for two non-equivalent molybdenum atoms in the enzyme active centre, with direct hydrogen transfer occurring to one of these only. However, in view of the above discussion, it is evident that incompleteness of exchange can be explained without this hypothesis. Thus, whilst direct hydrogen transfer seems well established, the relatively poor signal to noise ratios on the EPR traces obtainable at short reaction times has so far impeded work on this phenomenon. A further complication has arisen in using 1-methylxanthine for such work. This seemed a particularly suitable substrate, since it gives unusually fast development of the Rapid signal (54). However, multiple species involved in its Rapid signals appear to be different from those given by any other substrate and have not so far been fully studied. Hence, considerably more work seems to be called for on the direct hydrogen transfer phenomenon, to answer questions such as which of the several interacting protons involved in the Rapid signals is the one transferred from the substrate.

F. The Binding Site for Reducing Substrates and the Relationship between the Very Rapid and Rapid Signals

Evidence that substrates other than NADH react with the enzyme *via* its molybdenum now seems overwhelming. The most direct evidence for this in the case of purine substrates is the appearance of Mo(V) signals at an early stage in the catalytic cycle. With xanthine as substrate, the Very Rapid signal is the first signal observed (53), developing and reaching maximum intensity in a time which is quite short compared with the turnover. With 1-methylxanthine only the Rapid signal is observed but this comes up almost as quickly as does the Very Rapid signal for xanthine (54). Less direct evidence comes from the observations that complexing with purine substrates affects the molybdenum signals (Section V C) and that hydrogen atoms derived from purines are transferred in the catalytic reaction to a site interacting with molybdenum in the enzyme (Section V E).

If the EPR work may be thought conclusive in showing molybdenum to be the reaction site for purine substrates, then it is perhaps rather less direct and compelling for other substrates such as aldehydes. However, the data when supplemented by results obtained with the deflavo enzyme and in inhibition studies, leave little doubt that aldehydes do in fact react with molybdenum of the enzyme. The data relating to interaction with aldehydes will thus be summarized. Rapid freezing EPR studies of the 'single turnover type' (i. e. using small amounts of aldehyde in the presence of an excess of oxygen) have not been possible because of the high Km values for aldehydes, while approach-to-steady-state EPR experiments (59) were inconclusive since molybdenum, radical and iron-sulphur signals appeared with similar time-courses. Direct hydrogen transfer studies with deuterated salicylaldehyde (59, Section V E) although giving indications of behaviour similar to that seen with purines, were scarcely conclusive, presumably owing to slowness of the overall process relative to that of proton exchange with the medium. On the other hand, development of the Inhibited EPR signal when xanthine oxidase loses activity in the presence of formaldehyde (Section V A) or deuteroformaldehyde, has shown that this aldehyde can indeed become bound at a site which is clearly interacting with molybdenum and must be quite close to the metal (81). However, this information is not necessarily of direct relevance to the binding site for aldehydes in the catalytic reaction. Aside from EPR studies, work on deflavo xanthine oxidase has made it clear that flavin is not the binding site for aldehydes, as the deflavo enzyme is rapidly reduced by them in stopped-flow experiments (52). Additional evidence for this was adduced (52) from experiments in which cyanide rendered the enzyme non-reducible by aldehydes, reactivity towards NADH remaining unaffected. Although this work must be re-evaluated in the light of a later finding from the same laboratory (86), that this inhibitor reacts with sulphur in the enzyme and not with molybdenum, as had earlier been supposed (93; see Section V G), it nevertheless seems that the catalytic properties of the molybdenum site are affected by cyanide treatment whilst those of the iron-sulphur and flavin sites are not (as judged from the undiminished ability of the enzyme to catalyze NADH reduction of ferricyanide (52)).

Finally, in the case of inhibitory substrate analogues such as alloxanthine, strong evidence has recently been presented that these bind to molybdenum in reduced xanthine oxidase (33). If the enzyme is reduced with xanthine, then treated anaerobically with alloxanthine and finally exposed to air, catalytic activity is lost. Though flavin and iron in the final product are in the oxidized state, there are significant spectral differences between it and the native enzyme. These are believed (33) due to reduction of molybdenum from Mo(VI) to Mo(IV) and complexing of the latter with the inhibitor. This work is illustrated in Fig. 6 and represents the first good evidence for a contribution by molybdenum to the visible absorption spectrum of a molybdenum-containing enzyme.

In view of its importance (see Section V D), the evidence (33) that alloxanthine binds to enzyme with molybdenum reduced to Mo(IV) rather than to Mo(V) will be examined. The conclusion comes from measurements of the quantity of ferricyanide required to reverse alloxanthine inhibition. These experiments were complicated by reaction of the oxidizing agent with thiol groups, which necessitated working in the presence of a mercurial. Under suitable conditions, 1.6 equivalents of ferricyanide per mole of flavin were required for full re-activation. Correction for inactivated enzyme in the sample brought the value to 2.2 equivalents, implying (in the presumed absence of other redox active groups) oxidation of Mo(IV) to Mo(VI) (33). Although other investigators, using 2,6-dichlorophenol indophenol in place of ferricyanide, found

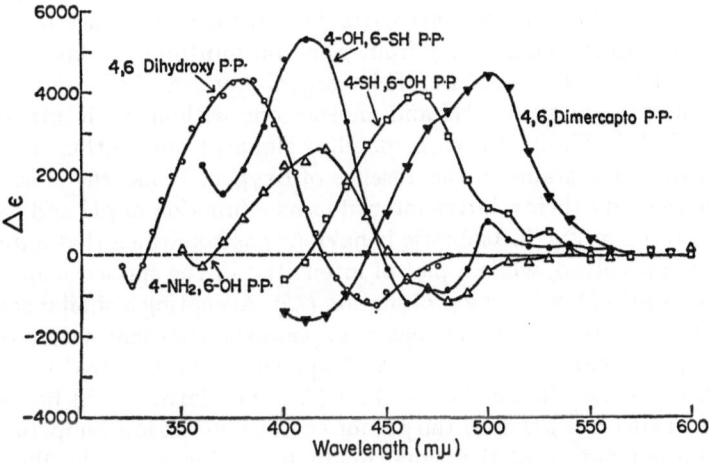

Fig. 6. Difference spectra between xanthine oxidase inactivated with various pyrazolo [3, 4-d] pyrimidines and the native enzyme. The spectra are believed to represent the increase in absorption occurring when Mo(VI) of native enzyme is converted to Mo(IV) complexed with the inhibitors. Spectra were obtained by treating the enzyme with inhibitors in the presence of xanthine, then admitting air, so as to re-oxidize the iron and flavin chromophores. The extinction coefficients, $\Delta \varepsilon$, are expressed per mole of enzyme flavin. Since some inactivated enzyme was present, extinction coefficients per atom of molybdenum of active enzyme will be about 30% higher than these values. (Reproduced from Ref. 33, with the permission of Dr. V. Massey.)

the number of oxidizing equivalents required to be less (*94*), it seems likely that this discrepancy was due to failure to correct for the presence of inactivated enzyme in the samples, rather than that reactivation involves Mo(V) → Mo(VI), as was proposed.

We now consider the relationship between the Very Rapid and the Rapid EPR signals. It is presumed that either can arise from the same molybdenum atom of the enzyme. With xanthine, at short reaction times, the Very Rapid signal predominates at high pH and the Rapid at low pH (*53*). The latter signal shows interaction with protons while the former does not. An obvious inference would be simply that the Very Rapid and Rapid signals are, respectively, high pH and low pH versions of the same intermediate, the difference between them representing no more than the dissociation at high pH of the interacting proton.

Though other explanations have been put forward (*87*), it now seems most likely that this is indeed the correct one. Difficulties of interpretation have arisen mainly because of the transient nature of the Very Rapid signal, which led to the idea (*76, 87*) that it might represent a precursor of the Rapid. In the explanation which we now favour the two signals are on essentially parallel reaction pathways, i.e. one is on a high pH path and the other on a low pH path. The strongest evidence favouring such a proposition comes, not from work on xanthine oxidase, but by analogy with sulphite oxidase. This enzyme gives a Mo(V) signal with proton interaction at low pH and another one without such interaction at high pH (*15*). Unlike the corresponding signals from xanthine oxidase, both of these are stable in the absence of oxygen. It has therefore been possible to study their relative intensities as a function of pH and results are illustrated in Fig. 7. Isobestic behaviour has confirmed that only two species are involved, while a plot of intensities of the species against pH indicates a pK of 8.2 for the protonation (*15*). Accepting a similar scheme for xanthine oxidase, one then has to explain the transient nature of the Very Rapid signal and the fact that it appears to be favoured when the enzyme turns over in the frozen state (*76*). The latter could be due to changes in effective pH or of the pK for protonation at low temperatures. The transient nature of the signal might be analogous to the Phase II 'fading' of the Rapid signal, i.e. due to reduction to Mo(IV). Another explanation, a variant of the 'precursor' scheme would be that the proton-accepting group exists in the oxidized enzyme but with a higher pK than it has in the reduced form, so that change from Very Rapid to Rapid signal in the early stages of reaction with xanthine would correspond to establishment of proton equilibrium in the system. Further work will be required to distinguish between these possibilities and also to explain the general absence of the Very Rapid signal with substrates other than xanthine.

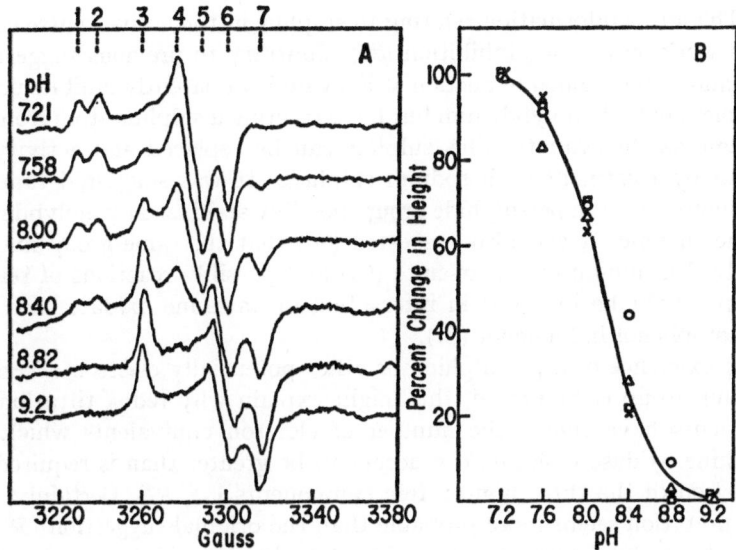

Fig. 7. Effect of pH on the EPR spectrum recorded at $-100°$ of sulphite oxidase reduced by sulphite. The species present at low pH values, which shows proton splitting, is replaced by another species at high pH. The pK for the transformation is about 8.2. In (A), maxima and minima in the derivative spectra are denoted by the numbers 1–7. In (B) changes in the spectra are plotted as a function of pH with values at pH 7.2 taken as 100% and those at pH 9.2 taken as 0%, or *vice versa*. The features in the spectra measured were: height of the 1 and 2 doublet (open circles); height of the peak at 3 (squares); distance between 4 and 5 (triangles) and height of 7 (diagonal crosses). (Reproduced from ref. *15*, with the permission of Dr. *K. V. Rajagopalan*.)

G. Evidence for Association of Sulphur-Containing Groups with Molybdenum

There is a considerable body of evidence in the literature favouring some type of association of molybdenum in the active centre of xanthine oxidase with one or more sulphur atoms. Apart from early speculations (*31*) based on reaction of the enzyme with mercurials, the first indications along these lines came from EPR data (*79*). It was suggested that relatively high g_{av} values and low molybdenum hyperfine couplings in signals from the enzyme were analogous to those in a series of Mo(V)-thiol complexes (*95*), indicating that the metal might have one or more sulphur ligands in the enzyme active centre. While the conclusion is quite likely correct, it may be added that high g_{av} values for Mo(V) compounds are not exclusive to those with sulphur ligands (see e.g. *96*).

135

The latest information relating to sulphur in the active centre comes from work on cyanide inhibition (86). Contrary to previous suggestions (93) and as mentioned in Section V F, cyanide apparently inhibits not by complexing with molybdenum but by extracting a sulphur atom from the enzyme as thiocyanate. This sulphur can be replaced and activity restored by treatment with sodium sulphide. It was suggested that the sulphur occurs as a persulphide group, possibly stabilized by molybdenum in the enzyme. It was also suggested (86) that the same group may be involved in inhibition by arsenite (93) and possibly reactions of related groups might be involved in interaction of xanthine oxidase with 2,6-dichlorophenol indophenol (94).

If existence of a persulphide or other potentially electron accepting sulphur group is confirmed, this might explain why redox titration experiments have shown the number of electron equivalents which the xanthine oxidase molecule can accept to be greater than is required for reduction of the three non-protein components (58, 91). Certainly, this interpretation seems more probable than the original suggestion (58, 91) that the molybdenum can be reduced to lower oxidation states than Mo(IV) by some substrates.

H. Conclusions

No doubt many aspects of the relation between structure and function in xanthine oxidase will not be understood until its structure can be determined by X-ray crystallography. The general conclusion from EPR work on the enzyme, is that the method has proved in this case to be a most revealing one for studying the processes of catalysis and that it has given a wealth of information of types not available from other techniques. However, EPR studies on their own leave large gaps in our knowledge. Therefore EPR work, even for an enzyme giving so many clearly defined signals as does xanthine oxidase, in order to be fully definitive, must be applied in conjunction with other more conventional methods.

A current overall picture of the reaction mechanism of xanthine oxidase, which differs substantially from one proposed earlier (87) is as follows. The enzyme is presumed to have two independent catalytic units, though this has not so far been proved rigorously. Reducing substrates are bound at molybdenum and reduce this from Mo(VI) both to Mo(V) and to Mo(IV). Reducing equivalents are then transferred by intramolecular reactions from molybdenum to iron-sulphur and also, either directly or via this, to flavin. Oxidizing substrates as a class, seem capable of reacting with all three types of centre in the enzyme. Thus, oxygen reacts predominantly with flavin, phenazine methosulphate reacts with molybdenum, and ferricyanide apparently reacts with iron-sulphur (52).

Further studies, particularly kinetic ones, will be required to elucidate the precise significance of complexes which substrate molecules form with molybdenum in the enzyme. Presumably such complexing may take place at the Mo(VI) and Mo(IV) levels as well as with Mo(V). Generally, the indications seem to be that Mo(V) may be of more direct significance in catalysis by the enzyme than is Mo(IV). Reduction of the metal to the five-valent level by the substrate implies the participation of substrate free radicals in the reaction though so far there is no direct evidence for this, possibly because any radicals formed disproportionate rapidly (67). Alternatively, there might be a two-electron transfer from the substrate with one of the electrons going to another grouping on the enzyme, such as the sulphur atom which seems to be necessary for a functional molybdenum. However, if this were the case, then the sulphur radical or other paramagnetic centre so produced ought to be formed simultaneously and in equal quantity with Mo(V) and there seems no reason why it should not be detectable by EPR. There is clearly no direct evidence favouring such a possibility. The first radical signal seen in rapid freezing EPR is from the flavin and it appears more slowly than the Mo(V) signal. The fate of the second reducing equivalent therefore remains a problem.

The overall reaction presumably involves the loss of two electrons and a proton from the substrate. The carbonium ion so produced would pick up a hydroxyl from the medium, to yield the product. In many cases the proton of the substrate is transferred directly to the enzyme where it is able to exchange with the solvent. However the apparent existence of a high pH reaction pathway in which the proton is not observed suggests that this may not be an obligatory step in catalysis. The actual nature of the proton accepting group might be an oxygen atom bound directly to molybdenum but there is no positive evidence for this as yet. It is interesting, though, that the proton involved in these processes exchanges quite slowly in comparison with normal exchange rates for protons bound to oxygen (97). Possibly hydrogen bonding is responsible for the slow exchange.

Further studies will be needed to establish the manner in which methanol and formaldehyde react in the region of molybdenum, thereby stabilizing the five-valent state of the metal and so inhibiting the enzyme. Further work will be needed, too, for an understanding of the exact nature of the reaction which is involved in a converting active xanthine oxidase to the inactivated form. Reaction at a neighbouring sulphur containing group which can act as a ligand to the molybdenum seems a strong possibility for both these processes.

VI. The Role of Molybdenum in other Enzymes

A. Sulphite Oxidase

Sulphite oxidase has been mentioned above (see Section V F). It may be noted that occurrence of molybdenum in this enzyme, which had not previously been suspected, was first established as a result of EPR work (15) and the observation thus represents one of the few recorded examples of EPR serving as an analytical method in biochemistry. The enzyme has been obtained in quite a high state of purity and the only non-protein constituent other than molybdenum is haem. The resting enzyme gives no molybdenum(V) EPR signals but these appear in the presence of the reducing substrate and disappear on reoxidation with ferricyanide or on further reduction with dithionite. The signals, as already noted, seem analogous to the Rapid and Very Rapid signal types from xanthine oxidase, that corresponding to the former having an exchangeable proton giving a splitting of 10—12 gauss. Though it seems that molybdenum is the enzyme constituent reacting with the substrate, sulphite, this has not yet been confirmed by rapid freezing EPR. There may be some form of association between sulphur and molybdenum in the enzyme since, like milk xanthine oxidase, it is inhibited by cyanide and by arsenite (15).

B. Xanthine Dehydrogenases and Aldehyde Oxidase

As mentioned earlier, the various xanthine dehydrogenases and aldehyde oxidase are all quite similar to milk xanthine oxidase. The reason for variation within this series in reactivity towards oxygen and other acceptors is not at present clear but is presumably not related to the molybdenum centre itself. Instead, it probably relates either to the flavin or more likely to the relative locations of the different functional units within the protein molecule as a whole. In this connection it is interesting to note that *Stirpe* and *Della Corte* (98) have reported that the xanthine dehydrogenase in rat liver scarcely reacts with oxygen if the enzyme is extracted under very mild conditions but that such extracts develop oxidase activity, subsequently, e.g. on storing at —20 °. It will be interesting to see if further work shows the changes occurring under these conditions to represent a modification in molecular geometry, e.g. molybdenum moving closer to flavin.

Xanthine dehydrogenase from chicken liver reacts readily with NAD as acceptor (17) while that from *Micrococcus lactilyticus* is inactive towards this, reacting instead with ferredoxin (18). Both enzymes react only slowly with oxygen. It seems reasonable to assume, however, that for each member of this group of enzymes, reducing substrates all react via molybdenum, as in milk xanthine oxidase. Presumably, different substrate specificities then represent differences in the environment of the molybdenum atoms within the active sites. Particular points of difference in the *Micrococcus* enzyme (*18*) are that it apparently fails to oxidize N-methyl nicotinamide and that it oxidizes purine only to the 8-hydroxy derivative, instead of to hypoxanthine, xanthine and uric acid. This enzyme is also reported to be effective in bringing about quite rapid dismutation of xanthine to hypoxanthine and uric acid. The substrate specificity of aldehyde oxidase is quite similar to that of xanthine oxidases and dehydrogenases. It oxidizes purine to 8-hydroxypurine and hypoxanthine to xanthine, as well as oxidizing aldehydes and N-methyl-nicotinamide (*99*).

Enzymes of this group show a number of similarities to one another. For instance, the chicken enzyme and aldehyde oxidase are both sensitive to methanol, to cyanide and to arsenite in much the same way as is milk xanthine oxidase (*93*). There are also similarities in EPR properties, though EPR studies on these enzymes have not been extensive since, compared to the milk enzyme, all are difficult to obtain in quantity. Thus, moderately detailed studies (*55, 100*) on aldehyde oxidase revealed a variety of molybdenum (V) EPR signals, depending on substrate and reaction time. No detailed analysis of the signals was attempted and proton interaction was not detected, though its existence seems by no means excluded. Possibly further study will show most of these signals to be analogous to the Rapid signals from milk xanthine oxidase. A particular point of interest with aldehyde oxidase is that a weak but clearly defined molybdenum signal is observed when the enzyme is in the resting state. This signal is somewhat reminiscent of the Slow signal from inactivated milk xanthine oxidase, particularly in that $g_\perp$ appears greater than $g_\parallel$. Though it may by tempting to assume existence of an analogous inactivated form of aldehyde oxidase, stable in air, some of the data (*55, 100*) cannot be readily reconciled with this hypothesis at present. Methanol treatment of aldehyde oxidase gives a signal strikingly similar to the Inhibited signal from the milk enzyme, except that the proton splittings (*100*) seem slightly larger than for xanthine oxidase (*87*). This observation must definitely confirm a basic similarity in the molybdenum active sites of the two enzymes.

Of other enzymes of the group, there seems to have been little EPR work on avian xanthine dehydrogenase (but see *93*) and only very limited work on the *Micrococcus* enzyme (*101*). Here the resting enzyme gives quite a strong molybdenum signal, which changes and decreases somewhat in the presence of substrate (*55, 101*). Clearly this is a significant area for further work since, if this finding is confirmed, a redox function for molybdenum in this enzyme could probably only be explained if important involvement of Mo(IV) in catalysis were assumed.

C. Nitrate Reductase

For nitrate reductase, evidence on the role of molybdenum in the catalytic mechanism of the enzyme from *Neurospora* was first presented in 1954 by *Nicholas* and *Nason* (*21*) and the position seems to have changed relatively little since then. The original conclusion (*23*) was that molybdenum functions as an electron carrier in the sequence:

$$NADPH \rightarrow Flavin \rightarrow Mo \rightarrow NO_3^-$$

with the metal alternating between the 5- and 6-valent forms. The methods used in this study were 'classical' ones (i.e. studies of the effects of added co-factors on native and molybdenum-free enzyme) and the difficulties involved in such work have been fully reviewed by *Singer* and *Massey* (*102*). Although a few EPR experiments (*25*) (on enzyme from another source) gave some indications of Mo(V) involvement and although work with non-enzymic model systems (e.g. *103*) appears consistent with the proposed scheme, it would clearly be desirable for detailed kinetic and EPR studies to be made on a nitrate reductase, in order to elucidate the mechanism fully. The overall reaction, $NO_3^- \rightarrow NO_2^-$, is of course a two-electron one but it can occur *via* the one-electron intermediate NO_2, since this product is capable (*103*) of disproportionating nonenzymically to give, ultimately nitrate plus nitrite. However, it would still be interesting to know whether Mo(V) and Mo(IV) are both involved in catalysis. It may be added that although milk xanthine oxidase is capable of utilizing nitrate as an oxidizing substrate (*30, 104*), the mechanism of this process does not seem to have been studied by modern methods.

D. Nitrogenase

Mechanistic studies on nitrogenase still seem to be in their infancy. For instance, there appears (cf. *114*) to be no firm evidence yet that the two protein components of the enzyme are involved in the form of a permanent complex, rather than associating and then dissociating again as part of the catalytic cycle. Though there have been speculations that molybdenum is at the binding site for nitrogen (cf. *105*), evidence favouring this (*111*) may still not be conclusive. The overall reaction involves transfer of six reducing equivalents to the nitrogen molecule to yield ammonia, no intermediates apparently being liberated during this reaction. Presumably a group of metal atoms in the active centre must all be involved in a concerted process to enable such an apparently improbable reaction to be achieved. It may be worth drawing an analogy with the copper-containing oxidases (*106, 107*) which reduce oxygen to water, a four-electron process, also achieved without liberation of intermediate products. Here, it seems that there are three different types of copper atoms in the enzyme molecule, each playing a different and quite specific role in the overall concerted process. One might therefore hope in nitrogenase to be able to distinguish different types of both iron and molybdenum atoms in the active centre. So far, however, nothing positive seems to have been established about the role of molybdenum in the enzyme. In fact, the EPR work to date (*7a, 5b, 108, 109, 110, 112*) has failed to reveal any evidence whatever for involvement of mononuclear Mo(V).

E. Conclusions

So little is known about molybdenum enzymes other than milk xanthine oxidase that there is little to be said by way of general conclusions. In all cases where there is direct evidence (except possibly for xanthine dehydrogenase from *Micrococcus lactilyticus*) it seems that molybdenum in the enzymes does have a redox function in catalysis. For the xanthine oxidases and dehydrogenases and for aldehyde oxidase, the metal is concerned in interaction of the enzymes with reducing substrates. However, for nitrate reductase it is apparently in interaction with the oxidizing substrate that the metal is involved. In nitrogenase the role of molybdenum is still quite uncertain.

Acknowledgements. We thank Dr. *H. Dalton* and Dr. *B. E. Smith* for helpful comments and Professor *J. R. Postgate*, Dr. *W. H. Orme-Johnson*, Dr. *V. Massey* and Dr. *J. F. Gibson* for allowing us to quote from unpublished work. The work was supported by grants from the Medical Research Council.

References

1. *Hewitt, E. J.:* Biol. Rev. *34*, 333 (1959).
2. *Anderson, A. J.:* In Inorganic Nitrogen Metabolism, p. 3 (Eds. *McElroy, W. D.,* and *Glass, B.*). Baltimore: Johns Hopkins Press 1956.
3. *Arnon, D. I., Stout, P. R.:* Plant Physiol. *14*, 599 (1939).
4. *Underwood, E. J.:* Trace elements in human and animal nutrition 2nd Edition p. 100. New York: Academic Press 1962.
5a. *Postgate, J. R.* (Ed.): The chemistry and biochemistry of nitrogen fixation. London: Plenum Press 1971.
5b. *Hardy, R. W. F., Burns, R. C., Parshall, G. W.:* Advances in Chemistry Series (Am. Chem. Soc.), No. 100, p. 219 (1971).
6. *Mortenson, L. E., Morris, J. A., Jeng, D. Y.:* Biochim. Biophys. Acta *141*, 516 (1967).
7a. *Dalton, H., Morris, J. A., Ward, M. A., Mortenson, L. E.:* Biochemistry *10*, 2066 (1971).
7b. *Vandercasteele, J.-P., Burris, R. J.:* J. Bacteriol. *101*, 794 (1970).

8. *Kelly, M., Lang, G.:* Biochim. Biophys. Acta *223*, 86 (1970).
9. *Cook, K. A., Eady, R., Kelly, M., Postgate, J. R., Smith, B. E.:* Unpublished and *Cook, K. A.:* D. Phil. Thesis, Univ. of Sussex 1971.
10. *Burns, R. C., Holsten, R. D., Hardy, R. W. F.:* Biochem. Biophys. Res. Commun. *39*, 90 (1970).
11. *Dervartanian, D. E., Bramlett, R.:* Biochim. Biophys. Acta *220*, 443 (1970).
12. *Taniguchi, S., Itagaki, E.:* Biochim. Biophys. Acta *44*, 263 (1960).
13. *Garrett, R. H., Nason, A.:* J. Biol. Chem. *244*, 2870 (1969).
14. *Downey, R. J.:* J. Bacteriol. *105*, 759 (1971).
15. *Cohen, H. J., Fridovich, I., Rajagopalan, K. V.:* J. Biol. Chem. *246*, 374 (1971).
16. *Rajagopalan, K. V., Fridovich, I., Handler, P.:* J. Biol. Chem. *237*, 922 (1962).
17. *— Handler, P.:* J. Biol. Chem. *242*, 4097 (1967).
18. *Smith, S. T., Rajagopalan, K. V., Handler, P.:* J. Biol. Chem. *242*, 4108 (1967).
19. *Hart, L. I., McGartoll, M. A., Chapman, H. R., Bray, R. C.:* Biochem. J. *116*, 851 (1970).
20. *McGartoll, M. A., Pick, F. M., Swann, J. C., Bray, R. C.:* Biochim. Biophys. Acta *212*, 523 (1970).
21. *Nicholas, D. J. D., Nason, A.:* J. Biol. Chem. *211*, 183 (1954).
22. *Knox, J. R., Prout, C. K.:* Chem. Commun. 1227 (1968).
23. *Nason, A.:* In The Enzymes 7, 587 (Eds. *Boyer, P. D., Lardy, H. A., Myrbäck, K.*). New York: Academic Press 1963.
24. *Lam, Y., Nicholas, D. J. D.:* Biochim. Biophys. Acta *178*, 225 (1969).
25. *Fewson, C. A., Nicholas, D. J. D.:* Biochim. Biophys. Acta *49*, 335 (1961).
26. *Cove, D. J., Coddington, A.:* Biochim. Biophys. Acta *110*, 312 (1965).
27. *Nason, A., Antoine, A. D., Ketchum, P. A., Frazier, W. A., Lee, D. K.:* Proc. Nat. Acad. Sci. *65*, 137 (1970).
28a. *Spence, J. T.:* Coordination Chem. Rev. *4*, 475 (1969).
28b. *Guest, J. R.:* Mol. Genet. *105*, 285 (1969).
29. *Cove, D. J.:* Proc. Roy. Soc. B *176*, 267 (1970).
30. *Ketchum, P. A., Cambier, H. Y., Frazier, W. A., Madansky, C. H., Nason, A.:* Proc. Nat. Acad. Sci. *66*, 1016 (1970).
31. *Bray, R. C.:* In: The Enzymes, 7, 533 (Eds. *Boyer, P. D., Lardy, H. A., Myrbäck, K.*). New York: Academic Press 1963.
32. *Knowles, P. F., Diebler, H.:* Trans. Faraday Soc. *64*, 977 (1968).
33. *Massey, V., Komai, H., Palmer, G., Elion, G. B.:* J. Biol. Chem. *245*, 2837 (1970).
34. *Bray, R. C.:* FEBS Letters *5*, 1 (1969).
35. *Ingram, D. J. E.:* Biological and biochemical applications of electron spin resonance. London: Adam Hilger 1969.
36. *Leigh, J. S.:* J. Chem. Phys. *52*, 2608 (1970).
37. *Bray, R. C.:* Biochem. J. *81*, 189 (1961).
38. *— Pettersson, R.:* Biochem. J. *81*, 194 (1961).
39. *Andrews, P., Bray, R. C., Edwards, P., Shooter, K. V.:* Biochem. J. *93*, 627 (1964).
40. *Greenlee, L., Handler, P.:* J. Biol. Chem. *239*, 1090 (1964).
41. *Fridovich, I.:* J. Biol. Chem. *241*, 3126 (1966).
42. *Mason, H. S.:* Science N. Y. *125*, 1185 (1957).
43. *Haddow, A., de Lamirande, G., Bergel, F., Bray, R. C., Gilbert, D. A.:* Nature *182*, 1144 (1958).
44. *Bray, R. C.:* Unpublished.
45. *Gibson, J. F., Bray, R. C.:* Biochim. Biophys. Acta *153*, 721 (1968).
46. *Ehrenberg, A., Bray, R. C.:* Arch. Biochem. Biophys. *109*, 199 (1965).

47. *Garbett, K., Gillard, R. D., Knowles, P. F., Strangroom, J. E.:* Nature *215,* 824 (1967).
48. *Johnson, C. E., Bray, R. C., Cammack, R., Hall, D. O.:* Proc. Nat. Acad. Sci. *63,* 1234 (1969).
49. *Palmer, G., Massey, V.:* J. Biol. Chem. *244,* 2614 (1969).
50. *Orme-Johnson, W. H., Beinert, H.:* Biochem. Biophys. Res. Commun. *36,* 337 (1969).
51a. *Hall, D. O., Evans, M. C. W.:* Nature *223,* 1342 (1969).
51b. *Tsibris, J. C. M., Woody, R. W.:* Coordination Chemistry Reviews *5,* 417 (1970).
52. *Komai, H., Massey, V., Palmer, G.:* J. Biol. Chem. *244,* 1692 (1969).
53. *Bray, R. C., Palmer, G., Beinert, H.:* J. Biol. Chem. *239,* 2667 (1964).
54. *— Knowles, P. F., Meriwether, L. S.:* In: Magnetic Resonance in Biological Systems, p. 249, (Eds. *Ehrenberg, A., Malmström, B. G., Vänngård, T.*). Oxford: Pergamon 1967.
55. *Rajagopalan, K. V., Handler, P., Palmer, G., Beinert, H.:* J. Biol. Chem. *243,* 3797 (1968).
56. *Knowles, P. F., Gibson, J. F., Pick, F. M., Bray, R. C.:* Biochem. J. *111,* 53 (1969).
57. *Orme-Johnson, W. H., Beinert, H.:* Biochem. Biophys. Res. Commun. *36,* 905 (1969).
58. *Massey, V., Brumby, P. E., Komai, H., Palmer, G.:* J. Biol. Chem. *244,* 1682 (1969).
59. *Bray, R. C.:* In: Flavins and Flavoproteins 3rd Int. Symp. p. 385 (Ed. *Kamin, H.*). Baltimore: University Park Press 1971.
60. *— Pick, F. M., Samuel, D.:* Europ. J. Biochem. *15,* 352 (1970).
61. *Fridovich, I.:* J. Biol. Chem. *245,* 4053 (1970).
62. *Nakamura, S., Yamazaki, I.:* Biochim. Biophys. Acta *189,* 29 (1969).
63. *Gutfreund, H., Sturtevant, J. M.:* Biochem. J. *73,* 1 (1959).
64. *Morell, D. B.:* Biochem. J. *51,* 657 (1952).
65. *Avis, P. G., Bergel, F., Bray, R. C.:* J. Chem. Soc. 1219 (1956).
66. *Bray, R. C., Pettersson, R., Ehrenberg, A.:* Biochem. J. *81,* 178 (1961).
67. *Swann, J. C., Bray, R. C.:* To be submitted to Europ. J. Biochem.
68. *Hart, L. I., Bray, R. C.:* Biochim. Biophys. Acta *146,* 611 (1967).
69. *Pawlik, R. T., Bray, R. C.:* Unpublished observations cited in reference 20.
70. *Nelson, C. A., Handler, P.:* J. Biol. Chem. *243,* 5368 (1968).
71. *Palmer, G., Bray, R. C., Beinert, H.:* J. Biol. Chem. *239,* 2657 (1964).
72. *McGartoll, M. A., Bray, R. C.:* Biochem. J. *114,* 443 (1969).
73. *Komai, H., Massey, V.:* In: Flavins and Flavoproteins 3rd Int. Symp. p. 399 (Ed. *Kamin, H.*). Baltimore: University Park Press 1971.
74. *Ball, E. G.:* J. Biol. Chem. *128,* 51 (1939).
75. *Bray, R. C., Malmström, B. G., Vänngård, T.:* Biochem. J. *73,* 193 (1959).
76. *Pick, F. M., Bray, R. C.:* Biochem. J. *114,* 735 (1969).
77. *Beinert, H.:* In: Flavins and Flavoproteins, 3rd Int. Symp. p. 416 (Ed. *Kamin, H.*). Baltimore: University Park Press 1971.
78. *Bray, R. C., Vänngård, T.:* Biochem. J. *114,* 725 (1969).
79. *— Meriwether, L. S.:* Nature *212,* 467 (1966).
80. *— Knowles, P. F., Pick, F. M., Vänngård, T.:* Biochem. J. *107,* 601 (1968).
81. *Pick, F., McGartoll, M. A., Bray, R. C.:* Europ. J. Biochem. *18,* 65 (1971).
82. *Maki, A. H., McGarvey, B. R.:* J. Chem. Phys. *29,* 35 (1958).
83. *Pick, F. M.:* Ph. D. thesis, University of London (1971).
84. *Massey, V.:* Personal communication.

85. *Bray, R. C., Chisholm, A. J., Hart, L. I., Meriwether, L. S., Watts, D. C.:* In: Flavins and flavoproteins p. 117 (Ed. *Slater, E. C.*). Amsterdam: Elsevier 1966.
86. *Massey, V., Edmondson, D.:* J. Biol. Chem. *245*, 6595 (1970).
87. *Bray, R. C., Knowles, P. F., Pick, F. M.:* Proc. Meet. Fed. Europ. Biochem. Soc. Prague, *16*, p. 267. New York: Academic Press 1968.
88. *Swann, J. C.:* Unpublished observations cited in reference 59.
89. *Bray, R. C., Swann, J. C.:* Unpublished.
90. *— Lowe, D. J., Pawlik, R. T.:* Unpublished.
91. *Massey, V., Komai, H., Palmer, G., Elion, G. B.:* Vitamins Hormones *28*, 505 (1970).
92. *Bray, R. C., Knowles, P. F.:* Proc. Roy. Soc. *A 302*, 351 (1968).
93. *Coughlan, M. P., Rajagopalan, K. V., Handler, P.:* J. Biol. Chem. *244*, 2658 (1969).
94. *Gurtoo, H. L., Johns, D. G.:* J. Biol. Chem. *246*, 286 (1971).
95. *Meriwether, L. S., Marzluff, W. F., Hodgson, W. G.:* Nature *212*, 465 (1966).
96. *Cousins, M., Green, M. L. H.:* J. Chem. Soc., *A 16* (1969).
97. *Amdur, I., Hammes, G. G.:* Chem. Kinetics: Principles and Selected Topics, p. 148. New York: McGraw-Hill 1966.
98. *Stirpe, F., Della Corte, E.:* J. Biol. Chem. *244*, 3855 (1969).
99. *Rajagopalan, K. V., Handler, P.:* J. Biol. Chem. *239*, 2027 (1964).
100. *— — Palmer, G., Beinert, H.:* J. Biol. Chem. *243*, 3784 (1968).
101. *Aleman, V., Smith, S. T., Rajagopalan, K. V., Handler, P.:* In: Flavins and Flavoproteins, p. 99 (Ed. *Slater, E. C.*). Amsterdam: Elsevier 1966.
102. *Singer, T. P., Massey, V.:* Record Chem. Progr. *18*, 201 (1957).
103. *Spence, J. T.:* Arch. Biochem. Biophys. *137*, 287 (1970).
104. *Westerfield, W. W., Richert, D. A., Higgins, E. S.:* In: Inorganic nitrogen metabolism p. 492 (Eds. *McElroy, W. D., Glass, B.*). Baltimore: Johns Hopkins Press 1956.
105. *Bui, P. T., Mortenson, L. E.:* Proc. Nat. Acad. Sci. *61*, 1021 (1968).
106. *Malkin, R., Malmström, B. G.:* Advan. Enzymol. *33*, 177 (1970).
107. *Brändén, R., Malmström, B. G., Vänngård, T.:* Europ. J. Biochem. *18*, 238 (1971).
108. *Kelly, M., Gibson, J. F.:* Unpublished.
109. *Smith, B. E., Bray, R. C.:* Unpublished.
110. *Evans, M. C. W., Telfer, A., Cammack, R., Smith, R. V.:* FEBS Letters *15*, 317 (1971).
111. *Burns, R. C., Fuchsman, W. H., Hardy, R. W. F.:* Biochem. Biophys. Res. Commun. *42*, 353 (1971).
112. *Davis, L. C., Shah, V. K., Brill, W. J., Orme-Johnson, W. H.:* Biochim. Biophys. Acta. In *press*.
113. *Forget, P.:* Europ. J. Biochem. *18*, 442 (1971).
114. *Silverstein, R., Bulen, W. A.:* Biochemistry 9, 3809 (1970).
115. *Spector, T., Johns, D. G.:* Biochem. Biophys. Res. Commun. *32*, 1039 (1968).

Received October 6, 1971

Evolution of Biological Iron Binding Centers*

J. B. Neilands

Department of Biochemistry, University of California, Berkeley, CA 94720, USA

Table of Contents

I. Introduction

A. General Significance of Iron in Biology

Among those elements essential for life, iron enjoys a status of extraordinary importance. It is involved in storage and transport of oxygen, in electron transport, in the metabolism of N_2 and H_2, in the reduction of ribotides to deoxyribotides (precursors of DNA), in oxidation and hydroxylation of a host of inorganic and organic metabolites and, finally, in the decomposition or utilization of hydrogen peroxide. In spite of its abundance in the earth's crust, the profound insolubility of the ferric ion at neutral pH has demanded the evolution of special ligands which can dissolve, transport and make available the element

---

* This review will not be concerned with functionally alternative structures and metabolites which appear in iron-limited growth. Thus *Clostridium pasteurianum* and other bacteria when grown in the presence of iron form ferredoxin; grown at low iron the same organisms form flavodoxin, a flavoprotein [*Knight, E., Jr., Hardy, R. W.*: J. Biol. Chem. *242*, 1370 (1967); *Mayhew, S. G., Massey, V.*: J. Biol. Chem. *244*, 794 (1969)].

to aerobic organisms. This affords yet another, for the most part relatively new, class of iron binding molecules the function of which is the transport of iron itself.

The number of identified chemical reactions in the single living species most studied from the biochemical viewpoint, *Escherichia coli*, is probably about a third of the total which can occur in this enteric organism and the genetic loci of many of these have been mapped on the chromosome (*1*). It will soon be possible to sum the metabolic iron of a cell such as *E. coli*, subtract this number from the total iron (as determined, for example, by atomic absorption spectroscopy) and so arrive at an estimate of the unidentified iron in the organism. Given a knowledge of all of the iron compounds in a cell it may be feasible to follow, by spectroscopic means, the flux of iron throughout the growth phases of the organism.

The particular metabolic role assigned to iron in biology will depend primarily upon the intrinsic physical and chemical properties of the element. However, as pointed out many years ago by *Theorell* (*2*), the atoms bound to the iron and its immediate environment will be of crucial significance in modulating the activity of the central metal ion. In the case of the hemoproteins, these are the so-called heme-linked groups. The latter determine the specificity of the protein and specify if it shall be hemoglobin, myoglobin, catalase, peroxidase, cytochrome, tryptophane pyrrolase, and so forth. This concept has lately been expanded to include groupings which are only indirectly in contact with the metal atom; although remotely located, such groupings must be of paramount importance in the reaction mechanism of the iron compounds.

Life, in any form, without iron is in all likelihood impossible (see Summary and Conclusions). Iron compounds can be expected to be encountered in all species from anaerobic bacteria to man. Accordingly, a comprehensive survey of the naturally occurring iron compounds must necessitate examination of the three broad kingdoms of living species, viz., microorganisms (protists), plants and animals.

The purpose of this paper will be to review the distribution of iron compounds in natural material. In regard to structure, emphasis will be placed on the atoms bonded directly to iron. However, for reasons just stated, such a superficial presentation is generally insufficient to explain the mechanism of action of the coordinated iron. We will not be concerned here with the specialized probes needed to obtain hints on the mode of binding of ferrous and ferric ions in macromolecules; such techniques have been described elsewhere *in extenso* (*3*).

Identification of the iron linked groups in those proteins not bearing the porphyrin nucleus has proven to be a formidable task and even relatively small protein molecules such as the ferredoxins have had to await a crystallographic solution to the structure of the iron center. Knowledge of the general properties of those molecules already structurally characterized should provide background information useful for further research into the nature of the iron atom in the host of metal conjugated proteins which exist in nature.

If iron is of crucial importance to life then, logically, the story of evolution can be related to inventions in the organic milieu which endowed the metal ion with an ever-increasing array of functional abilities. Of the total different iron compounds extracted from living material, the largest group belongs to those assigned the task of microbial iron transport, i.e., siderochromes (4). Since these molecules are typically 1000 daltons or less in molecular weight, they have been regarded as not sufficiently "semantic" to provide much evidence about evolution (5). However, since they are produced generally by aerobic microorganisms they may serve as a relatively fine means of calibration of development within narrow ranges of species. Unlike the non-porphyrin iron proteins, the siderochromes are amenable to characterization up to the level of solution conformation (6) and the atoms ligated to the iron can generally be assigned with confidence. This series might well disclose the full range of complexing abilities for ferric ion which is available to living cells.

B. Precisely Known Physiological Roles of Iron

Throughout the vertebrate animal kingdom, the hemoproteins myoglobin and hemoglobin are involved in oxygen storage and transport. Proteins of closely related structure occur in the root nodules of N_2 fixing plants and in yeasts; the function of these molecules is not completely understood (7). Some invertebrate animals also use a heme-type nucleus for oxygen transport, i.e., in the giant molecules termed erythrocruorins and chlorocruorins (8); other invertebrates, such as *Polychaetes*, *Sipunculoids*, *Priapuloids*, *Brachiopodes* and various marine worms utilize for a similar purpose the non-heme iron containing protein hemerythrin (9). Iron protoporphyrin IX, the most ubiquitous of the iron porphyrins, also serves as the prosthetic group of the hydroperoxidases (10—12). Historically, catalase has been associated with animals and peroxidase with plants; however, both types of these hydrogen-peroxide metabolizing enzymes are now recognized as widely distributed throughout the phyla.

Hemoproteins which engage in electron transport — the cytochromes — are much more widely dispersed among living species and occur in microorganisms, plants and animals (13). Again there are two types of iron proteins which can perform the task of electron transport, the heme and the non-heme. The latter term has become practically synonymous

147

in usage with the designation "iron-sulfur" proteins and this is to be regretted since sulfur is not always the ligand in these molecules (cf. transferrin). The natural distribution of the iron sulfur proteins ranges through the microbial, plant and animal kingdoms (14).

Oxygenation and hydroxylation of a wide variety of biological materials almost always involves the participation of a metal ion, usually iron and sometimes copper. In one unique case, tryptophane pyrrolase, the iron is present as heme (15). The only dioxygenase enzyme reaction in which a metal ion has not been implicated is one involved in the degradation of vitamin B_6 (16).

A component of the ribotide reductase complex of enzymes, protein B_2, has been shown to contain two non-heme iron atoms per mole (17). This enzyme plays a vital, albeit indirect, role in the synthesis of DNA. Curiously, the lactic acid bacteria do not employ iron for the reduction of the 2' hydroxyl group of ribonucleotides. In these organisms this role has been assumed by the cobalt-containing vitamin B_{12} coenzyme (18). The mechanism of the reaction has been studied and has been shown to procede with retention of configuration (19).

In addition to these more-or-less well characterized proteins, iron is known to be bound to certain flavoproteins such as succinic dehydrogenase (20), aldehyde oxidase (21), xanthine oxidase (22) and dihydroorotate dehydrogenase (23). Iron is present and functional in non-heme segments of the electron transport chain but again no real structural information is at hand (24).

Iron transport agents may belong to the protein or non-protein class. In the former group are found the animal proteins transferrin (25), lactoferrin (26) and conalbumin (27). The low molecular weight iron carrying compounds from microorganisms, the siderochromes, may occur with or without a bound metal ion. Typically, severe repression of biosynthesis of these substances can be expected to set in at an iron concentration of ca. 2×10^{-5} g atoms/liter (28). Most, but not all, of these substances can be described as phenolates or hydroxamates (4).

Finally, animal, plant and microbial tissues have been shown to contain the iron storage protein ferritin. The animal protein has been extensively studied, but the mechanism of iron binding has not been completely resolved (29). Animal tissues contain, in addition, a type of granule comprised of iron hydroxide, polysaccharide and protein. The latter, called hemosiderin, may represent a depository of excess iron (30). Interestingly, a protein with properties parallel to those of ferritin has been found in a mold. Here the function of the molecule can be examined with the powerful tools of biochemical genetics (31).

Iron appears to play a significant role in the mechanism of action of aconitase, a constituent enzyme of the tricarboxylic acid cycle (32).

The two non-proteinaceous substances ferroverdin (*33*) and pyrimine (*34*) are unique in that they preferentially form ferrous complexes. A number of iron-containing antibiotics have been isolated (*35*) but of these only a few such as albomycin (*36*) and ferrimycin (*37*) have been characterized.

Table 1 lists the major iron-containing substances known in biology at the present time.

Table 1. *Classification of some iron containing substances of biological origin*[a])

Substance	Source
Oxygen transport and storage proteins	
Heme proteins	
Hemoglobin	Vertebrates
Leghemoglobin	Root nodules
Erythrocruorin	Invertebrates
Chlorocruorin	Invertebrates
Myoglobin	Vertebrates
Non-heme proteins	
Hemerythrin	Annelids
Hydroperoxidases	
Catalase	Animals (commonly)
Peroxidase	Plants (commonly)
Cytochrome *c* peroxidase	Yeast
Electron transport proteins	
Heme proteins	
Cytochromes	Microorganisms, plants, animals
Non-heme proteins	
Ribotide reductase protein B$_2$	*E. coli*, animal tissues
Ferredoxins	Bacteria, plants
Rubredoxins	Bacterial
Adrenodoxins	Adrenal gland
Putidaredoxin	Bacterial
High potential iron protein (HIPIP)	Photosynthetic bacteria
Related iron-sulfur proteins	Bacterial (commonly)
Nitrogenase	
Molybdo-iron protein	Bacterial
Iron protein	Bacterial
Hydrogenase	Bacterial

[a]) References to the literature may be found in the text. For a list of iron-containing metalloproteins and metalloenzymes see Ref. (*38*). The classification shown is arbitrary since in some cases the function must be regarded as inferrred rather than demonstrated.

149

J. B. Neilands

Table 1 (continued)

Substance	Source
Oxygenases and hydroxylases	
Heme proteins	
Tryptophane dioxygenase	Bacteria, plants, animals
Non-heme proteins	
Lysine dioxygenase	Bacterial
Catechol dioxygenase	Bacterial
Related oxygenases	Diverse sources
Iron flavoproteins	
Succinic dehydrogenase	Coincident with TCA cycle
Aldehyde oxidase	Porcine liver
Xanthine oxidase	Milk
Dihydro orotate dehydrogenase	*Zymobacterium oroticum*
Iron transfer and storage substances	
Protein	
Transferrin	Blood
Lactoferrin	Milk
Conalbumin	Egg white
Ferritin	Animal tissues, fungi, plants
Hemosiderin	Animals
Non-protein (siderochromes)	
Phenolates	Enteric bacteria
Hydroxamates	Bacteria, yeasts, fungi
Miscellaneous iron compounds	
Protein	
Aconitase	Coincident with TCA cycle
Agavain	Sisal extract
Non-protein	
Ferroverdin	Bacterial
Pyrimine	Bacterial

II. Relevant Biogeochemical Properties of Iron

At 35 g/100 g iron is the major element in the whole Earth. Although surpassed in the Earth's crust by oxygen and silicon here again its level of 5 percent is exceeded only, among the metals, by aluminum. The concentration in sea water is 10^{-6} to 10^{-7} gram atoms/liter, a value generally considered too low to accomodate maximum growth rates for very aerobic microorganisms. Iron and nickel constitute the core

150

of the Earth and since the former metal occurs commonly in meteorites it is believed to be universally distributed throughout the solar system.

It is now well accepted that the atmosphere of the primitive Earth consisted of water vapor, methane, ammonia and molecular hydrogen. In general, the elements, including iron, were believed to be in the most reduced state. With energy furnished by ultraviolet and high energy radiation, chemical evolution proceeded apace (39). Accepting the age of the Earth as $\sim 4.5 \times 10^9$ years, the anaerobic phase could have persisted some billion years since fossil remains of algae have been detected in rocks dated at ca. $\sim 1 \times 10^9$ years (40, 41). Since the solubility product of Fe (OH)$_3$ is less than 10^{-37}M it is obvious that ligating systems had to be devised in order to maintain an effective level of soluble ferric ion throughout the millenia of anaerobic evolution in which the ferrous-ferric couple was doubtless called upon to perform electron transfer.

A salient feature of iron and one which is absolutely crucial to its biological role is its ability to exist in two stable valence states. Depending on the nature of the ligand, the redox potential of the Fe^{II}/Fe^{III} couple may vary over a wide range. Increasing unsaturation in the binding species tends to favor the ferrous state; this is a consequence of both the opportunity of π bonding in such complexes and the higher electron pressure in ferrous iron. Ferric ion has low affinity for amine ligands but is strongly bound to oxygen atoms as these occur, for instance, in phenolates, β-diketones and hydroxamates. The common coordination numbers for both ferric and ferrous ions are 4 and 6. Most complexes are octahedral but tetrahedral iron is known in both the inorganic and biological spheres. *Malmström* has reviewed some of the basic coordination properties of iron such as, for example, the spin state in strong and weak ligand fields (42). These properties have afforded the collection of some information on the environment of iron in macromolecules by application of probes such as electron spin resonance and Mössbauer spectroscopy (3). Generally, these techniques can only rule out certain configurations.

III. Iron Compounds in Protist Organisms

A. Anaerobic Bacteria

Certain strictly anaerobic bacteria and lactic acid bacteria apparently do not contain heme compounds. In the first named organisms this cannot be ascribed to a failure to perform the first step in porphyrin biosynthesis since *Clostridia* are notorious for production of the porphyrin-like nucleus (corrin) which occurs in vitamin B$_{12}$ (7, 43).

151

Do these prokaryotic organisms display a primitive iron metabolism, in the sense that they form a relatively limited number of iron compounds, and are they accordingly good candidates to suspect as direct descendants of the primordial cell?

Although heme is absent in *Clostridia*, it was early recognized that anaerobic bacteria may contain substantial levels of iron (*44*). To date the best characterized iron compounds from this source are the iron-sulfur proteins termed ferredoxins and rubredoxins. Molecular structures of representatives of both types of protein have been worked out by *Jensen* and his colleagues by X-ray diffraction analysis (see below).

Ferrodoxin-like proteins are characterized by the generation of H_2S following treatment with dilute HCl. Molecules of this type are probably best designated as proteins rather than as "enzymes", the latter term indicating much greater specificity. They are widely distributed in living cells and are found in aerobic bacteria, the free living nitrogen-fixing bacteria (*Azotobacter*), photosynthetic bacteria (*Chromatium*), chloroplasts and even the tissues of higher animals (adrenal gland). Their function is that of redox catalysts in steroid 11-β-hydroxylation (adrenodoxin), camphor hydroxylation (putidaredoxin), photoevolution of O_2 (chloroplast ferredoxin), and the phosphoroclastic reaction of pyruvate (*clostridial* ferredoxin) (*14*).

Although also iron-sulfur proteins, the rubredoxins do not generate H_2S on acidification since in this case the thiol groups are contributed by cysteinyl residues in the polypeptide chain. The function of clostridial rubredoxin is as yet unknown; in *Pseudomonas* sp. a similar protein catalyzes the ω-hydroxylation of alkanes, a reaction requiring molecular O_2.

While the oxidation reduction potential of the ferredoxins is -0.2 V to -0.4 V and that of the rubredoxins is about -0.05 V, a protein from the photosynthetic bacterium *Chromatium* has a redox potential of $+0.35$ V. This is the high potential iron protein, or HIPIP.

All members of the class of iron-sulfur proteins contain, per mole, from 1 to 8 iron atoms and from zero (rubredoxins) to 8 labile sulfides.

The crystallographic structure of rubredoxin from *Clostridium pasteurianum* at 2.5 Å, a resolution sufficient to reveal the sequence of several of the bulky amino acid side chains, shows the iron coordinated to two pairs of cysteine residues located rather near the termini of the polypeptide chain (Fig. 1). A related rubredoxin, with a three times larger molecular weight, from *Pseudomonas oleovorans* is believed to bind iron in a similar fashion. This conclusion is based on physical probes, especially electron paramagnetic resonance spectroscopy, all of which indicate that the iron is in each case situated in a highly similar environment; however, the proteins display some specificity in catalytic function.

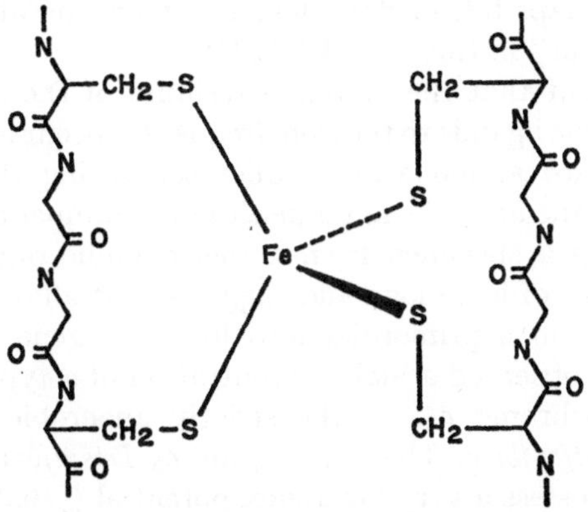

Fig. 1. The iron-cysteinyl center of *Clostridium pasteurianum* rubredoxin (*45*). Two centers of this type are proposed to occur in *Pseudomonas oleovarans* rubredoxin (*46*)

Recently *Jensen* and co-workers have determined the structure of a clostridial-type ferredoxin obtained from *Micrococcus aerogenes* (*47*). One of the two apparently identical iron-sulfur clusters is illustrated in Fig. 2. The structure is compatible with a model with iron and labile sulfide at alternate corners of a cube. This accounts for the equivalence of these moieties in the protein. Another 8-iron-8 labile sulfur ferredoxin, from *Clostridium acidiurici*, similarly contains two independent iron-sulfur clusters per molecule (*48*). *Strahs* and *Kraut* (*49*) had earlier discovered

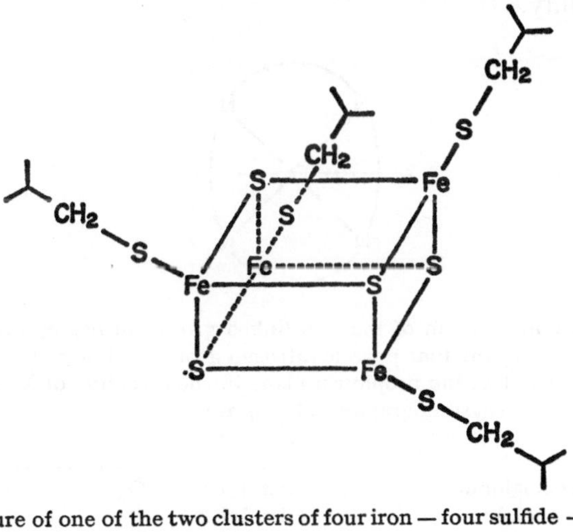

Fig. 2. Structure of one of the two clusters of four iron — four sulfide — four cysteinyl residues of *Micrococcal* ferredoxin (*47*)

"a compact, perhaps tetrahedral cluster of four iron atoms as the most likely arrangement" in the case of HIPIP.

It is significant that the cysteinyl residues of these proteins, which are presumably the ligands to the iron, frequently occur in a 1—4 sequence, viz. there are two amino acid residues separating the cysteine pair. Investigation of the amino acid sequences of a number of bacterial ferredoxins, including a specimen from a thermophilic organism, reveals a substantial degree of homology and suggests that all of these iron-sulfur proteins arose from a primordial ferredoxin by gene duplication (50).

Postgate (51) observed a high concentration of a type of cytochrome, designated cytochrome c_3, in the strictly anaerobic sulfate-reducing bacterium, *Desulfovibrio*. The cytochrome c_3 *Desulfovibrio vulgaris* has been shown to possess a very low redox potential (—0.205 V), two heme groups per mole and a sequence which implies two half sections, each of which binds a heme group (52).

Although oxygen was found to be the only oxidant for conversion of coproporphyrinogen III to protoporphyrin IX, anaerobic systems must obviously exist for the biosynthesis of the latter molecule (43). Porphine itself has not been found in nature but spectral lines identical to those of bis-pyridylmagnesiumtetrabenzoporphine have been detected in interstellar space (53).

Fig. 3 illustrates schematically the structures of some well characterized hemoproteins. In the cytochrome series the desired range of redox potentials is achieved by variations in the axial ligands, contributed by the porphyrin moiety, as well as by substitutions around the periphery of the tetrapyrrole nucleus with groups of differing electron-attracting ability.

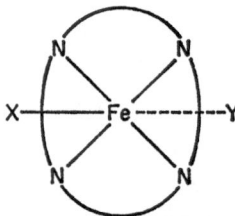

Fig. 3. Schematic illustration of the iron linked groups in hemoproteins. The porphyrin ring containing the four pyrrole nitrogen atoms is shown as an elipsoid. The iron atom is 0.5 Å out of the prophyrin plane in the direction of X (54). The axial ligands confirmed by crystallography (55, 56) are:

	X	Y
Oxyhemoglobin	Histidyl N	O_2
Oxymyoglobin	Histidyl N	O_2
Fe^{II} or Fe^{III} Cytochrome c	Histidyl N	Methionyl sulfur

B. Organisms Metabolizing Primitive Earth Reactants

In tracing the evolutionary development of iron ligands it is of interest to examine the machinery employed by organisms which carry out reactions on those substances believed to have been present on the primitive Earth. Specific substrates acted on by this group include, besides ferrous iron itself, hydrogen sulfide, hydrogen gas, methane and reduced nitrogen compounds. Species which perform photosynthesis may be presumed to have the capacity to synthesize protoporphyrin IX since this substance is an intermediate in chlorophyll biosynthesis (43).

Non-photosynthetic microorganisms deriving energy from oxidation of ferrous iron are mainly of the heterotrophic group. However, the biochemistry and physiology of the obligate chemoautotrophes *Gallionella* sp. and *Ferrobacillus ferrooxidans* (*Thiobacillus ferrooxidans*) have been studied to some extent. Interest in these species has accelerated in view of the fact that the reaction initiated with ferrous sulfate,

$$4\ FeSO_4 + O_2 + 10\ H_2O \rightleftharpoons 4\ Fe(OH)_3 + 4\ H_2SO_4$$

results in a water pollution problem, viz., acid mine waters (57). *Ferrobacillus ferrooxidans* has been shown to contain cytochrome c (58, 59).

The colorless sulfur-oxidizing microorganisms include relatives of the bluegreen algae (*Beggiatoa* and *Thiothrix*) and a genus of the eubacteria (*Thiobacillus*). The latter require iron for growth and presumably synthesize the cytochrome systems (60).

The capacity to utilize hydrogen gas as a source of energy and/or reducing power is common to several types of eubacteria, including photosynthetic and non-photosynthetic species, pseudomonads, sulfate-reducers, enteric bacteria, micrococci, strict anaerobes and, finally, the free-living nitrogen-fixing *Azotobacter* (61). In the activation of molecular hydrogen by the reaction

$$H_2 \rightleftharpoons 2\ H$$

those hydrogen bacteria belonging to the genus *Hydrogenomonas* use oxygen as the electron acceptor. The group is of special interest in the present context since it represents an instance of side-by-side occurrence of both iron-sulfur proteins and hemoproteins within the same species. A similar situation exists for the facultative photoanaerobe, *Rhodospirillum rubrum*, which contains both cytochrome c_2 and ferredoxin (62).

Fang and *Burris* (63) isolated a c-type cytochrome from the cells of *Hydrogenomonas eutropha*. Type c cytochromes have been isolated from the cells of a number of other bacterial species (64) and a three dimensional model has been proposed for the cytochrome c_{551} of *Pseudomonas aeru-*

155

ginosa (*65*). *Hydrogenomonas eutropha* yielded one acidic and two basic cytochromes. The former was found to have an amino acid composition not substantially different from that of other bacterial cytochromes (*63*).

Hydrogenase and other components of the N_2 fixing apparatus of bacteria have been shown to be non-heme iron proteins (*66*).

The cytochrome system of enzymes has long been known to be associated with the activity of the nitrifying bacteria which catalyze, for example, the oxidation of ammonia:

$$NH_3 + 1^{1/2} O_2 \rightleftharpoons HNO_2 + H_2O$$

Organisms which oxidize methane exclusively and use this substrate as a source of carbon and energy appear to be restricted to *Methanomonas* sp. The latter is a strict aerobe which presumably contains cytochromes (*67*).

In summary, the capacity to synthesize both hemoproteins and iron sulfur proteins appears to be a ubiquitous attribute in organisms attacking reduced inorganic substrates. If the reactions by which these cells obtain energy represent relics of ancient forms of metabolism, it can only be concluded that heme formation and iron-sulfur coordination must have been invented at a very early stage in evolution.

C. Microbial Iron Transport Compounds (Siderochromes)

This topic has been reviewed recently (*4*, *68*) and the present discussion can be limited to aspects of the distribution and evolution of the siderochromes. The latter are here defined as specific ligands synthesized by microorganisms for the purpose of incorporating iron. The criteria for preliminary assignment of a natural product to this class should be (a) the formation is repressed by iron, (b) the compound exhibits a large differential in its avidity for ferric vs. ferrous iron, the former being greatly favored, and (c) mutants unable to synthesize either the particular ligand or its "permease" are defective in iron uptake. Of the siderochromes listed in Table 2, only enterobactin (enterochelin) and ferrichrome have thus far been shown to satisfy most of these criteria. However, these are the only two members of the class which have been studied in any detail.

The siderochromes recorded in Table 2 constitute a selected group and represent only about half of the total number of these substances isolated to date. However, these representatives display sufficient diversity in structure and in distribution to provide clues about their phylogenetic relationships.

Table 2. *Some siderochromes of protist organisms*[a])

Compound	Ligand System		Source
	Type	Number/mole	
Enterobactin (enterochelin)	Catechol	3	*Aerobacter aerogenes,Escherichia coli, Salmonella typhimurium*
2-N, 6-N-Di-(2,3-di-hydroxybenzoyl)-L-lysine	Catechol	2	*Azotobacter vinelandii*
2,3-Dihydroxy-N-benzoyl-L-serine	Catechol	1	*Aerobacter aerogenes,Escherichia coli, Salmonella typhimurium*
2,3-Dihydroxyben-zoyl-glycine (itoic acid)	Catechol	1	*Bacillus subtilis*
Mycobactins	Hydroxamic acid	2	*Mycobacterium smegmatis*
	Phenolic hydroxyl	1	*M. tuberculosis*
	Tertiary N	1	*M. kansasii*
			M. balnei
			M. piscium
			M. thamnopheos
			M. aureum
			M. fortuitum
			M. thermoresistibile
			M. marinum
			M. phlei
			M. terrae
Aerobactin	Hydroxamic acid	2	*Aerobacter aerogenes*
	Carboxylic acid	1	
	Alcoholic hydroxyl	1	
Schizokinen	Hydroxamic acid	2	*Bacillus megaterium*
	Carboxylic acid	1	
	Alcoholic hydroxyl	1	
Ferrichromes	Hydroxamic acid	3	Species of *Aspergillus, Neurospora, Paecilomyces, Penicillium, Spicaria, Ustilago, Cryptococcus, Actinomyces, Streptomyces* and probably *Sphacelotheca*
Rhodotorulic acid	Hydroxamic acid	2	Species of *Rhodotorula, Sporobolomyces, Sporidiobolus, Leucosporidium* and *Rhodosporidium*
Dimerum acid	Hydroxamic acid	2	*Fusarium dimerum*
Coprogen	Hydroxamic acid	3	*Penicillium* sp. and *Neurospora crassa*
Fusarinines	Hydroxamic acid	1–3	Species of *Fusaria, Aspergillus, Gibberella,* and *Penicillium*
Ferrioxamines	Hydroxamic acid	3	Species of *Nocardia, Micromonospora, Streptomyces,* and *Actinomyces*

[a]) For a more complete list of microbial siderochromes and for structural formulae see Ref. (*4*).

157

In view of their common function it is clearly desireable to describe the microbial iron transport compounds under a common name, viz., siderochrome (or equivalent), even though the latter designation was originally applied to the pure hydroxamate type ligand (69). Thus "siderochrome" was a chemical classification and it made no commitment about the biological role of the substances grouped under this title. The compounds are members of a family wherein the iron binding atoms may be furnished by three hydroxamic acid groups (ferrichromes, ferrioxamines); two hydroxamic acid groups, a phenolic hydroxyl and a tertiary nitrogen atom (mycobactins); two hydroxamic acid groups, a carboxylic acid group and an alcoholic hydroxyl (aerobactin, schizokinen); and, finally, one, two or three catechol residues (itoic acid, α-ε-bis-2,3-dihydroxy-benzoyl-lysine, enterobactin, respectively).

The prototype of the catechol siderochromes is enterobactin (70) [enterochelin in the terminology of Australian workers (71)]. Enterobactin occurs commonly throughout the enteric bacteria and enterobactin (enterochelin) is the natural carrier invoked when the organisms are

placed under low-iron stress, i. e., when iron is withheld from the medium or when certain additives make iron unavailable. Lower homologues of enterobactin, such as dihydroxybenzoyl serine, are apparently degradation products of the trimer but certain of these monomeric units may serve in some cases to transport iron, albeit less efficiently.

The catechol-type ligand appears to be restricted to siderochromes derived from prokaryotic microorganisms. *Klebsiella oxytoca*, an organism closely related to members of the genus *Aerobacter*, forms the 2,3-dihydroxy-N-benzoyl derivates of serine and threonine in three day cultures (72). It is not known if the latter amino acid occurs in trimers but examination of space-filling CPK models does indicate that enterobactin could accomodate a methyl substituent on the β carbon of the serine residue. Catechols occur in higher protist organisms but their formation in these species has not been intimately connected to iron transport (73).

It is interesting to note that the dihydroxybenzoyl nucleus arises from chorismic acid which, in turn, is derived from erythrose phosphate and phosphoenol pyruvate, both of these substances being intermediates in the anaerobic metabolism of carbohydrate (74). Accordingly, the biogenesis of the catechol type ligand is independent of the presence of oxygen gas.

The ability to form spores is an important development in the differentiation of eubacteria and it is perhaps not surprising that, in the siderochrome series, the genus *Bacillus* is a common ground for both catechol and hydroxamic type ligands. Thus while *Aerobacter* sp. form both enterobactin and aerobactin (75), *Bacillus megaterium* produces schizokinen (76) and *Bacillus subtilis* elaborates itoic acid (77). In the case of both aerobactin and schizokinen, the hydroxamic acid groups are erected on a platform consisting of citric acid. The latter is an intermediate in carbohydrate metabolism which, in contrast to erythrose and pyruvic acid, lies closer to oxygen. It is tempting to speculate that organisms electing a more aerobic way of life have assigned to citrate derivatives the duty of iron transport.

The mycobactins contain, like aerobactin and schizokinen, only two hydroxamic acid groups, the third chelate ring being formed by condensation of the nitrogen and oxygen atoms of serine or threonine with a salicyl moiety (78). The mycobactins have thus retained from the catechol series a lone phenolic hydroxyl and the same two hydroxy amino acids, serine and threonine, which occur in the enteric siderochromes. The mycobacteria and the actinomycetes, while equipped with a typical prokaryotic internal organization, resemble the fungi in their ability to form a mycelial vegetative construction (67).

Further evidence that the hydroxamic-acid containing compounds represent a more highly evolved type of siderochrome is given by inspection of their mode of biosynthesis which, in every case that has been critically examined, requires molecular oxygen (68, 79).

As seen from Table 2, the ferrichrome type of siderochrome is widely scattered throughout the *Ascomycetes*, *Basidiomycetes* and *Fungi Imperfecti*. It can almost be assumed, *a priori*, that all members of these groups form siderochromes. Serine and glycine, as in the catechol type siderochromes, once more commonly serve the function of building up the platform to which the iron binding atoms are attached. In all of these hexapeptides the iron is coordinated to a tripeptide of acyl-δ-N-hydroxy-L-ornithine; the remaining tripeptide is comprised of glycine, serine or alanine. The presence of the former two amino acids in the catechol type siderochromes has already been noted. Given the opportunity for further variation in the nature of the acyl substituent, a very large number of derivatives of the basic ligand structure are possible and many have been already discovered in the higher protists.

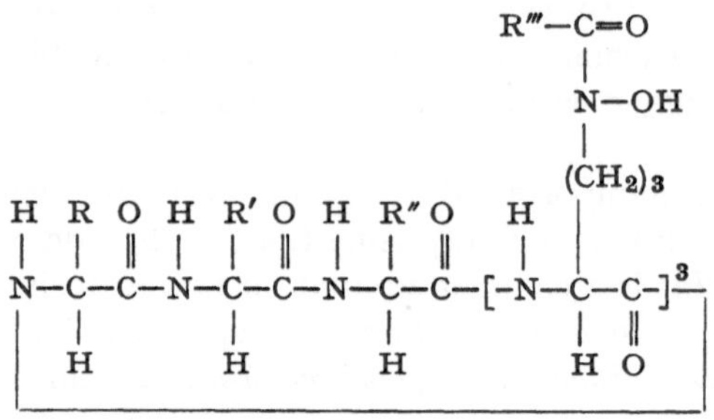

$$R'''{-}C{=}O$$
$$|$$
$$N{-}OH$$
$$|$$
$$(CH_2)_3$$

ligand system of ferrichrome type siderochromes

| H | R | O | H | R' | O | H | R'' | O | H |
| | | ‖ | | | ‖ | | | ‖ | |
N—C—C—N—C—C—N—C—C—[—N—C—C—]³—
| | | | | | |
H | H | H | H O

The lack of structural "conservatism" in the siderochromes, as contrasted to cytochrome c, can be illustrated by comparing representatives of these two iron-containing compounds from two adjacent classes under the phylum *Eumycophyta*, the *Ascomycetes* and the *Basidiomycetes*.

Coprogen, the siderochrome of the *Ascomycete, Neurospora crassa*, consists of a cyclic dipeptide of δ-N-hydroxyornithine acylated on one side by the ω-alcohol analogue of β-methyl glutaconic acid and on the other side by a residue of N-acetyl fusarinine (*80*). Obviously, coprogen differs radically in constitution from ferrichrome, from the *Basidiomycete, Ustilago sphaerogena*, in which R=R'=R''=H and R'''=CH₃ (see structure of the ferrichrome ligand).

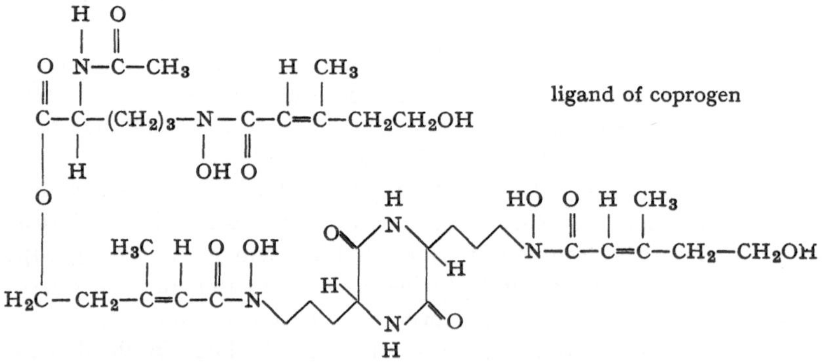

ligand of coprogen

A complete amino acid sequence for cytochrome c from *N. crassa* is available (*81*). In the case of *U. sphaerogena*, only an amino acid composition has been determined. *U. sphaerogena* forms a relatively large amount of cytochrome c when the organism is grown in its sporidial stage in certain media (82). The hemoprotein has been extracted at alkaline pH, purified by chromatography on Amberlite IRC-50, and some of its properties have been reported (83). Subsequently it was found that the cytochrome could be readily extracted at neutral pH from an acetone powder of the cells. By a slight modification of the general method (84) for purification of cytochrome c, the protein was obtained in crystalline form in a yield of about 10 mg/50 g of acetone powder (85). This preparation, the first to be obtained in the crystalline state from a member of the Basidiomycetes, is of interest from the point of view of evolutionary and comparative biochemistry. Despite the fact that the isoelectric point of the protein obtained by alkaline extraction of U. sphaerogena is several pH units below that of the classical cytochrome c of mammalian heart muscle it functions quite efficiently in the succinic dehydrogenase-cytochrome oxidase system in washed mitochondria. Even though the low isoelectric point of the fungal cytochrome may be an artifact of isolation, i.e., hydrolysis of amide, it may define the charge requirement for function in the assay system[1]).

The hydrodynamic properties of Ustilago cytochrome c were investigated by Thelander (86). He found the partial specific volume to be 0.721 ml/g and the molecular weight, by sedimentation equilibrium, to be 15,500. The latter value, although higher than that given by summation of the constituent amino acid residues (i.e., 11,877, see Table 3), indicates that the protein is monomeric.

The data presented in Table 3, which includes the amino acid composition of baker's yeast and Candida krusei cytochrome c for comparison, show that Ustilago and Neurospora cytochrome c contain the same number of total residues. In seven instances, the number of residues of a particular amino acid/mole are identical. Thus, even in the absence of a sequence for the Ustilago cytochrome it can be concluded that this protein, unlike the siderochromes, has suffered little alteration in the progression from the Ascomycetes to the Basidiomycetes. This can be ascribed to the varying function of the two types of molecules. Cytochrome c must fit into a relatively specific slot bounded by a reductase and an oxidase and it has hence evolved much more slowly than the more freely acting transport agents where the specificity constraints are less demanding.

In spite of the profound diversity in siderochrome structure observed throughout the kingdom Protista, some connecting links are clearly discernible. In addition to the already noted frequent use of the small

---

[1]) According to Komai (85) crystalline Ustilago cytochrome c isolated by cell fracture rather than by alkaline extraction also displays a neutral isoelectric point.

J. B. Neilands

Table 3. *Amino acid composition of cytochrome c from some protist organisms*

Amino acid	Ustilago sphaerogena	Neurospora crassa[a]	Saccharomyces cerevisiae[a]	Candida kreusei[a]
Asx	10	13	11	8
Thr	9	9	8	7
Ser	3	3	4	6
Glx	12	8	9	10
Pro	4	3	4	7
Gly	13	15	12	12
Ala	7	9	7	12
Cys	2	2	3	2
Val	5	1	3	3
Met	1	2	2	3
Ile	3	5	4	3
Leu	9	7	8	6
Tyr	4	4	5	5
Phe	6	6	4	4
Lys	11	14	16	12
His	2	2	4	4
Arg	5	3	3	4
Trp	1	1	1	1
Total residues	107	107	108	109
Identities with *Ustilago*		7	3	3

[a]) See Ref. (*81*).

neutral amino acids glycine and serine [alanine also occurs in some ferrichromes (*4*)], and the hydroxy amino acids serine and threonine, the ω-N-hydroxy derivatives of ornithine and lysine occur routinely in the hydroxylamino moiety of hydroxamic acid type siderochromes. Cyclic dipeptides of the decca-[ferribactin (*87*)], hexa-[ferrichrome (*4*)], and die-[rhodotorulic acid (*88*), dimerum acid (*80*)] peptide class contain δ-N-hydroxyornithine. Decarboxylated forms of the N-hydroxy amino acids are polymerized into the ferrioxamines. In a number of instances, α-β unsaturation occurs in the acyl substituent of the hydroxamic acid type siderochrome. Ester or amide bonds are employed to polymerize the monomeric units in both catechol and hydroxamic acid types of substance and where ester bonds are present in trimers esterases develop in aging cultures to degrade trimers back to monomers. Finally, *Salmonella typhimurium*, even though it is not known to synthesize hydroxamate type ligands, does nonetheless contain a cell bound uptake system or "permease" for ferrichrome and related compounds (*89*). Some

162

attempts have already been made to utilize for taxonomic purposes the characteristic distribution of hydroxamic acid type siderochromes within species (*78, 90*).

Recently, a new natural product linking sulfur, oxygen and iron has been isolated from bacterial sources. These substances, the thiohydroxamic acids, are worthy of discussion in the present context even though their biological significance has yet to be defined. The copper and iron complexes of N-methyl thioformylhydroxamic acid were obtained from the culture broth of *Pseudomonas fluorescens* and named fluopsins C and

F, respectively (*91*). In a practically simultaneous report, also from Japan, it was noted that the same ligand, named thioformin, could be secured by removal of copper from the antibiotic YC73, the latter produced by a "pseudomonad" culture (*92*). Fluopsins C and F are potent antibiotics and display broad spectrum activity against Gram positive and negative strains (*93*). The biosynthesis of N-methyl thioformylhydroxamic acid does not appear to be repressed by iron and *n*-paraffin is the preferred carbon source. In a comprehensive review on the chemistry of the thiohydroxamic acids the properties of the metal complexes were noted, including the fact that S/O coordination has been confirmed by crystallography (*94*). Biologists should be quite interested in the redox potentials and relative affinities for transition metal ions of compounds of this type.

IV. Iron Compounds in Plants

Plants appear to contain few, if any, of the general types of iron compounds which are not represented elsewhere in the protist or animal kingdoms.

Peroxidase activity has long been associated with extracts of plant tissue and the crystalline enzyme from horse radish root has been studied *in extenso*, particularly in regard to its mechanism of action (*11*). Plants also contain ferredoxin and various specialized cytochromes, both of which substances play an essential role in photosynthesis (*95, 96*). Agavain, a crystalline proteolytic enzyme from the leaves of *Agave*, contains 0.11% iron, a value corresponding to a minimum molecular weight of 53,000 daltons (97). Root nodule cells, as already noted, may contain leghemoglobin if the cells harbor the bacteroids of *Rhizobium* sp. (7). The cytochromes *c* of a number of higher plants have been sequenced and these data, in view of the absence of an adequate fossil record, should provide information on the origin of these species (98).

V. Iron Compounds in Animals

Reference to Table 1 indicates that the iron compounds which are unique to animals are those involved in the transport of oxygen and iron. This is a consequence of the higher level of differentiation in the animal organism, a development which precludes simple diffusion as a means of supplying the cell with essential nutrilites.

A complete crystallographic structure has been measured for myoglobin and hemoglobin (55). Most of the very high molecular weight oxygen-carrying proteins of invertebrates also contain iron protoporphyrin IX, but in the case of chlorocruorin from polychaete worms the prosthetic group is a formyl-substituted porphyrin (43).

The hemerythrin of *Golfingia gouldii* consists of eight subunits, each of which contains two iron atoms, in a protein with molecular weight 108,000. Spectral and magnetic data point to an oxo-bridged structure around the non-heme iron atom (99). Protein B_2 of ribotide reductase of *E. coli* has some properties in common with hemerythrin; presumably a protein corresponding to that of *E. coli* reduces ribotides in animal tissues, a conclusion based on probes with inhibitors.

The iron binding center of transferrin has been thought to consist of two histidyl residues, three tyrosine hydroxyls and a bicarbonate ion (25). However, the question should be regarded as still open since *Aisen et al.* (100) showed that the ferric complex could be formed by substituting other ligands for HCO_3^-. If tyrosyl hydroxyls are indeed linked to the iron this would constitute an interesting parallel with the siderochromes of eubacteria where aromatic hydroxyls constitute a common ligand. Proteins very similar to transferrin occur in milk (26) and in egg white (27) but, as in the case of transferrin, the structure around the iron atoms, of which there are two/mole, is again not completely understood.

The core of the iron storage protein ferritin consists of a hydrated ferric oxide-phosphate complex. Various models have been proposed which feature $Fe^{III} O_6$ oct., $Fe^{III} O_4$ tet. or $Fe^{III} O_4$ tet. $Fe^{III} O_6$ oct. complexing; the first listed is preferred by *Gray* (99) on the basis of the electronic absorption spectrum. The protein very closely related to ferritin which occurs in the mold *Phycomyces blakesleeanus* contains about half as much iron per mole as mammalian ferritin. It is localized in crystalline monolayers on the surface of liquid droplets (*31*). In plants ferritin is found in monolayers in proplastids (*101*).

The phosphoprotein phosvitin may attach a large number of ferric ions/mole, according to *Gray* (*99*) [citing work of *Saltman* and *Multani*], the number bound may be as large as **46**. Ferric ion, although firmly complexed, may be removed from this egg yolk protein by dialysis against solutions of ethylenediamine tetraacetic acid (*102*).

VI. Summary and Conclusions

It is apparent from this cursory survey that, while lacking the charisma of DNA, iron has played a prominent role in the evolution and development of living forms. One of the few, perhaps the only, species which can challenge the dictum that life and iron are inseparable is the group collectively called the lactic acid bacteria. In 1947 *MacLeod* and *Snell* (*103*) examined the mineral requirements of representative strains of these organisms. They reported:

"In no case did omission of iron from the medium affect growth deleteriously. Pretreatment of the medium with *Saccharomyces carlsbergensis* reduced the amount of iron to a level slightly below that required for optimum growth of this organism, but all of the lactic acid bacteria grew optimally on this medium without addition of iron. From the known facts that lactic acid bacteria grow anaerobically, do not contain cytochrome, and are catalase negative, it would be expected that if iron is required at all, it would be required in extremely small amounts." The highly sensitive *Csaky* (*104*) test for bound hydroxylamine is negative for latic acid bacteria (*105*), indicating a complete absence of the hydroxamate type of siderochromes. All of these observations lead to the conclusion that the lactic acid bacteria, which require many vitamins and amino acids, represent a type of regressive physiological adaptation. The species may have evolved in special environments, such as in the oral cavity, where most of its multifarious nutritional needs, with the exception of iron, were in ready supply. It is thus tempting to speculate that the microbes learned to live without iron entirely and, as a consequence, adopted a variant, non-iron means of forming deoxyribotides, viz., vitamin B_{12} coenzyme.

The remarkable range of redox potentials in the iron sulfur proteins, already noted, illustrates the principle that nature, having discovered a ligand system, attempts to extract from it the maximum utility. Certainly the outstanding problem awaiting solution in these proteins is an explanation for the relationship between structure and redox potential. Thus the potential of HIPIP is some 0.7 V higher than that of clostridial ferredoxin, even though both proteins contain very similar iron sulfur clusters. And chloroplast ferredoxin, with only two iron- two labile sulfide/mole, displays EPR spectra and redox potentials which are very similar to those of the ferredoxin from anaerobic bacteria (47).

Clearly, the all-sulfur type ligand constitutes a very effective means of complexing both Fe^{II} and Fe^{III} in both aerobic and anaerobic protoplasm.

Sulfur is utilized also in cytochrome c, via methionine 80, for the sixth ligand to iron. The elegant nuclear magnetic resonance spectroscopic measurements of *Gupta* and *Redfield* (106) demonstrate conclusively that the methionyl sulfur atom remains within the coordination sphere of the metal atom in both ferri- and ferro-cytochrome c. Sulfur and iron come in proximity again in the ribotide reductase system where electrons are donated by reduced thioredoxin, the latter a low molecular weight protein containing a pair of cysteine residues in the sequence cys-X-Y-cys. Iron is a component of the catalytic subunit, protein B_2, which transfers electrons from the reduced thioredoxin to the oxygen at the 2^1 position in the ribose moiety of the nucleoside diphosphates.

The thiohydroxamic acids of *Pseudomonas* sp present a unique and highly interesting example of S and O coordination. The discovery of these substances confirms the axiom that all possible types of iron-binding systems can be expected to be found in microorganisms. It has long been assumed (107) that iron must be reduced in order to be liberated from hydroxamate type siderochromes and *Komai* (85) found that dialyzed broken cells of *Ustilago* can catalyze reduction of ferrichrome iron by TPNH. However, the biological significance of this reaction is unknown. Recently, *Ratledge* (108) demonstrated a DPNH dependent reduction of mycobactin-ferric complex in mycobacteria. This reductase might be a likely target for the thiohydroxamic acid antibiotics.

All-oxygen chelation appears to be confined to the transport and storage molecules, i.e., in substances which are supposed to be rather highly evolved. Paradoxically, this type of coordination, as exemplified by the binding of hydroxyl ion, must have been the first which iron encountered. *Spiro* and *Saltman* (109) have reviewed the chemical, physical and biological properties of inorganic polymers of iron hydroxide. They found citrate, at concentrations < 20 mM, to be effective in promoting polynuclear complex formation. Such complexes are not the means of depriving enterobactin-less mutants of *Salmonella* of iron, however, since this organism apparently cannot utilize iron-citrate monomers.

Since the early work of *Warner* and *Weber* (110) it has been assumed that the alcoholate and carboxylate groups of citrate complex to ferric ion. This needs to be confirmed in the schizokinen-aerobactin series by some means of direct analysis since the fourth proton released from the iron complexes of these molecules could arise from ionization of a bound water molecule.

Regarding the siderochromes it is perhaps noteworthy that species bearing mitochondria appear to favor production of the hydroxamate type of ligand. The manner of mitochondrial uptake of iron from the cytoplasm is obscure, but the system must be an active one in view of the high concentration of iron proteins in these organelles. If mitochondria arose from ingested bacteria which became, in turn, endosymbionts, one would expect a permease for iron to survive in the mitochrondrial envelope. Mammalian tissues, at least under normal circumstances, do not appear to form hydroxylamines[2]). The feeding of amino fluorenes leads to N-hydroxylation and an increased level of carcinogenicity (*111*). It is perhaps for the latter reason that hydroxamates, which experienced a florid proliferation as the higher protist organisms struggled to protect their iron supply in an atmosphere containing a poisonous new gas — oxygen — have been rejected as iron carriers in animals.

In summary, the fragmentary evidence at hand suggests that sulfur was the primordial iron complexing material and that this ligand system was selectively modified by stepwise transition to a mixed sulfur-nitrogen and sulfur-oxygen and, finally, to an all-oxygen type of coordination.

Acknowledgment. The author is indebted to *A. Baldesten* for performing amino acid analyses on *Ustilago* cytochrome *c*.

VII. References

1. *Watson, J. D.:* Molecular Biology of the Gene. New York: W. A. Benjamin 1970, Second Edition.
2. *Theorell, H.:* Advan. Enzymol. 7, 265 (1947).
3. *Bearden, A. J., Dunham, W. R.:* Struct. Bonding 8, 1 (1970).
4. *Neilands, J. B.:* In: Inorganic Biochemistry; edited by *Eichhorn, G.* New York: Elsevier 1972.
5. *Zuckerkandl, E., Pauling, L.:* J. Theoret. Biol. 8, 357 (1965).
6. *Llinás, M., Klein, M. P., Neilands, J. B.:* J. Mol. Biol. 52, 399 (1970).
7. *Lascelles, June:* Tetrapyrrole Biosynthesis and Its Regulation. New York: W. A. Benjamin 1964.
8. *Lehmann, H., Hunstman, R. G.:* Man's Haemoglobins. Amsterdam: North Holland 1966.

---

[2]) It has recently been claimed that peptide-bound N-hydroxy amino acids occur in human brain tumor, in virus-induced tumor of mouse spleen, and in carcinogen-induced tumors in the rat (*112*).

9. *Ghiretti, F.:* In: Oxygenases, p. 517; edited by *Hayaishi, O.* New York: Academic Press 1962.
10. *Nicholls, P., Schonbaum, G. R.:* In: The Enzymes, p. 147; edited by *Boyer, P. D., Lardy, H. A., Myrbäck, K.* New York: Academic Press 1964.
11. *Saunders, B. D., Holmes-Siedle, A. G., Stark, B. P.:* Peroxidase. Washington: Butterworths 1964.
12. *Yonetani, T., Ohnishi, T.:* J. Biol. Chem. *241*, 2983 (1966).
13. *Okunuki, K., Kamen, M. D., Sekuzi, I.:* Structure and Function of Cytcchromes. Baltimore: University Park Press 1968.
14. *Rabinowitz, J. C.:* In: Bioinorganic Chemistry, p. 322; edited by *Dessy, R., Dillard, J., Taylor, L.* Washington: American Chemical Society 1971.
15. *Hayaishi, O., Nozaki, M.:* Science *164*, 389 (1969).
16. *Sparrow, L. G., Ho, P. P. K., Sundaram, T. K., Zach, D., Nyns, E. J., Snell, E. E.:* J. Biol. Chem. *244*, 2590 (1964).
17. *Brown, N. C., Eliasson, R., Reichard, P., Thelander, L.:* Biochem. Biophys. Res. Commun. *30*, 522 (1968).
18. *Blakeley, R. L., Chambeer, R. K., Nixon, P. F., Vitols, E.:* Biochem. Biophys. Res. Commun. *20*, 439 (1965).
19. *Fraser-Reid, B., Radatus, B.:* J. Am. Chem. Soc. *93*, 6342 (1971).
20. *Kearney, E. B.:* J. Biol. Chem. *235*, 865 (1960.)
21. *Rajagopalan, K. V., Fridovich, I., Handler, P.:* J. Biol. Chem. *237*, 922 (1962).
22. *Massey, V., Brumby, P. E., Komai, H., Palmer, G.:* J. Biol. Chem. *244*, 1682 (1969).
23. *Aleman, V., Handler, P.:* J. Biol. Chem. *242*, 4087 (1967).
24. *Hatefi, Y., Stempel, K. E., Hanstein, W. G.:* J. Biol. Chem. *244*, 2358 (1969).
25. *Bezkorovainy, A., Grohlich, D.:* Biochem. J. *123*, 125 (1971).
26. *Groves, M. L.:* In: Milk Proteins, p. 368; edited by *McKenzie, H. L.* New York: Academic Press 1971.
27. *Feeney, R. E., Komatsu, S. K.:* Struct. Bonding *1*, 149 (1966).
28. *Garibaldi, J. A., Neilands, J. B.:* Nature *177*, 526 (1956).
29. *Bryce, C. F. A., Crichton, R. R.:* J. Biol. Chem. *246*, 4198 (1971).
30. *Weinfeld, A.:* In: Iron Deficiency, p. 329; edited by *Hallberg, L., Harwerth, H.-G., Yannotti, A.* New York: Academic Press 1970.
31. *David, C. N., Easterbrook, K.:* J. Cell. Biol. *48*, 15 (1971).
32. *Villafranca, J. J., Mildvan, A. S.:* J. Biol. Chem. *246*, 5791 (1971).
33. *Candeloro, S., Grdenic, D., Taylor, N., Thompson, B., Niswamitra, M., Crawfoot-Hodgkin, D.:* Nature *224*, 589 (1969).
34. *Shiman, R., Neilands, J. B.:* Biochemistry *4*, 2233 (1965).
35. *Umezawa, H.:* Index of Antibiotics from Actinomycetales. State College, Pa.: University Park Press 1967.
36. *Mikes, O., Turkova, J.:* Chem. Listy *58*, 65 (1964).
37. *Bickel, H., Mertens, P., Prelog, V., Seibl. J., Walser, A.:* In: Anti-microbial Agents and Chemotherapy, p. 1951; edited by *Hobby, G. L.* Ann Arbor, Mich.: American Society for Microbiology 1965.
38. *Vallee, B. L., Wacker, W. E. C.:* In: Handbook of Biochemistry, p. C-51; edited by *Sober, H. A.* Cleveland, Ohio: Chemical Rubber Co. 1970, Second Edition.
39. *Wald, G.:* Proc. Natl. Acad. Sci. *52*, 595 (1964).
40. *Kenyon, D. H., Steinman, G.:* Biochemical Predestination. New York: McGraw Hill 1969.
41. *Singer, T. P.:* In: Biochemical Evolution and the Origin of Life, p. 203; edited by *Schoffeniels, E.* Amsterdam: North Holland 1971.

42. *Malmström, B. G.:* In: Iron Deficiency, p. 9; edited by *Hallberg, L., Harwerth, H.-G., Vannotti, A.* New York: Academic Press 1970.
43. *Marks, G. S.:* Heme and Chlorophyll. London: Van Nostrand 1969.
44. *Curran, H. R., Brunstetter, B. C., Myers, A. T.:* J. Bacteriol. *45*, 485 (1943).
45. *Herriott, J. R., Sieker, L. C., Jensen, L. H., Lovenberg, W.:* J. Mol. Biol. *50*, 391 (1970).
46. *Lode, E. T., Coon, M. J.:* J. Biol. Chem. *246*, 791 (1970).
47. *Sieker, L. C., Adman, E., Jensen, L. H.:* Nature *235*, 40 (1972).
48. *Orme-Johnson, W. H., Sweet, R. M., Sundaralingan, M., Dahl, L. F., Beinert, H.:* personal communication.
49. *Strahs, G., Kraut, J.:* J. Mol. Biol. *35*, 503 (1968).
50. *Tanaka, M., Haniv, M., Matsueda, G., Yasunobu, K. T., Himes, R. H., Akagi, J. M., Barnes, E. M., Devanathan, T.:* J. Biol. Chem. *246*, 3953 (1971).
51. *Postgate, J. R.:* J. Gen. Microbiol. *14*, 545 (1956).
52. *Ambler, R. P.:* Biochem. J. *109*, 47 (1968).
53. *Johnson, F. M.:* Chem. Eng. News, November *1*, 5 (1971).
54. *Koenig, D. F.:* Acta Cryst. *18*, 663 (1965).
55. *Dickerson, R. E., Geis, I.:* The Structure and Action of Proteins. New York: Harper and Row 1969.
56. — J. Mol. Evol. *1*, 26 (1971).
57. *Schnaitman, C. A., Korczynski, M. S., Lundgren, D. G.:* J. Bacteriol. *99*, 552 (1969).
58. *Blaylock, B., Nason, A.:* J. Biol. Chem. *238*, 3453 (1963).
59. *Din, G. A., Suzuki, I., Lees, H.:* Can. J. Microbiol. *45*, 1523 (1967).
60. *Lees, H.:* Biochemistry of Autotrophic Bacteria. London: Butterworths 1965.
61. *Stanier, R. Y., Doudoroff, M., Adelberg, E. A.:* The Microbial World. Englewood Cliffs, N. J.: Prentice-Hall 1963, Second Edition.
62. *Dus, K., Sletten, K.:* In: Structure and Function of Cytochromes, p. 293; edited by *Okunuki, K., Kamen, M. D., Sekuzu, I.* Baltimore: University Park Press 1968.
63. *Fang, F. S., Burris, R. H.:* J. Bacteriol. *96*, 298 (1968).
64. *Yamanaka, T., Okunuki, K.:* J. Biol. Chem. *239*, 1813 (1964).
65. *Dickerson, R. E.:* J. Mol. Biol. *57*, 1 (1971).
66. *Nakos, G., Mortenson, L. E.:* Biochemistry *10*, 2442 (1971).
67. *Horio, T., Kamen, M. D.:* Ann. Rev. Microbiol. *24*, 399 (1970).
68. *Emery, T.:* Advan. Enzymol. *35*, 135 (1971).
69. *Prelog, V.:* In: Iron Metabolism, p. 73; edited by *Gross, A.* Berlin–Göttingen–Heidelberg–New York: Springer 1964.
70. *Pollack, J. R., Neilands, J. B.:* Biochem. Biophys. Res. Commun. *38*, 989 (1970).
71. *Cox, G. B., Gibson, F., Luke, R. K. J., Newton, N. A., O'Brien, I. G., Rosenberg, H.:* J. Bacteriol. *104*, 219 (1970).
72. *Korth, H.:* Arch. Mikrobiol. *70*, 297 (1970).
73. *Ratledge, C., Chaudhry, M. A.:* J. Gen. Microbiol. *66*, 71 (1971).
74. *Gibson, F., Pittard, J.:* Bacteriol. Rev. *42*, 465 (1968).
75. — *Magrath, D. I.,* Biochem. Biophys. Acta *192*, 175 (1969).
76. *Mullis, K. B., Pollack, J. R., Neilands, J. B.:* Biochemistry *10*, 4894 (1971).
77. *Ito, T., Neilands, J. B.:* J. Am. Chem. Soc. *80*, 4645 (1958).
78. *Snow, G. A.:* Bacteriol. Rev. *34*, 99 (1970).
79. *Akers, H., Llinás, M., Neilands, J. B.:* Manuscript in preparation.
80. *Keller-Schierlein, W., Diekmann, H.:* Helv. Chim. Acta *53*, 2035 (1970). — *Diekmann, H.:* Arch. Mikrobiol. *73*, 65 (1970).

81. *Dayhoff, M. O.:* Atlas of Protein Sequence and Structure. Silver Spring, Md.: National Biomedical Research Foundation 1969.
82. *Grimm, P. W., Allen, P. J.:* Plant Physiol. *29*, 369 (1954).
83. *Neilands, J. B.:* J. Biol. Chem. *197*, 701 (1952).
84. *Hagihara, B., Morikawa, I., Tagawa, K., Okunuki, K.:* Biochem. Prep. *6*, 1 (1958).
85. *Komai, H.:* Doctoral Dissertation, University of California, Berkeley, 1968.
86. *Thelander, L.:* Personal communication.
87. *Maurer, B., Müller, A., Keller-Schierlein, W., Zähner, H.:* Arch. Mikrobiol. *60*, 326 (1968).
88. *Atkin, C. L., Neilands, J. B.:* Biochemistry 7, 3734 (1968).
89. *Pollack, J. R., Ames, B. N., Neilands, J. B.:* Federation Abstr. *29*, 3132 (1970).
90. *Atkin, C. L., Neilands, J. B., Phaff, H.:* J. Bacteriol. *103*, 722 (1970).
91. *Shirahata, K., Deguchi, T., Hayashi, T., Matsubara, I., Suzuki, T.:* Antibiotics (Tokyo) *23*, 546 (1970).
92. *Egawa, Y., Umino, K., Ito, Y., Okuda, T.:* J. Antibiotics (Tokyo) *24*, 124 (1971).
93. *Itoh, S., Inuzuka, K., Suzuki, T.:* J. Antibiotics (Tokyo) *23*, 542 (1970).
94. *Walter, W., Schaumann, E.:* Synthesis, p. 111, No. 3 March (1971).
95. *Knaff, D. K., Arnon, D. I.:* Proc. Natl. Acad. Sci. *64*, 715 (1969).
96. *Bishop, N. I.:* Ann. Rev. Biochem. *40*, 197 (1971).
97. *Tipton, K. F.:* Biochim. Biophys. Acta *110*, 415 (1965).
98. *Thompson, E. W., Notton, B. A., Richardson, M., Boulter, D.:* Biochem. J. *124*, 787 (1971).
99. *Gray, H. B.:* In: Bioinorganic Chemistry, p. 365; edited by *Dessy, R., Dillard, J., Taylor, L. Washington*: American Chemical Society 1971.
100. *Aisen, P., Aasa, R., Malmström, B. G., Vänngård, T.:* J. Biol. Chem. *242*, 2484 (1967).
101. *Robards, A. W., Humpherson, P. G.:* Planta *76*, 169 (1967).
102. *Taborsky, G.:* Biochemistry 2, 266 (1963).
103. *MacLeod, R. A., Snell, E. E.:* J. Biol. Chem. *170*, 351 (1947).
104. *Csáky, T. Z.:* Acta Chem. Scand. 2, 450 (1948).
105. *Burnham, B. F., Neilands, J. B.:* J. Biol. Chem. *236*, 554 (1961).
106. *Gupta, R. K., Redfield, A.:* Biochem. Biophys. Res. Commun. *41*, 273 (1970).
107. *Neilands, J. B.:* Bacteriol. Rev. *21*, 101 (1957).
108. *Ratledge, C.:* Biochem. Biophys. Res. Commun. *45*, 856 (1971).
109. *Spiro, T. G., Saltman, P.:* Struct. Bonding *6*, 116 (1969).
110. *Warner, R. C., Weber, I.:* J. Am. Chem. Soc. *76*, 2111 (1954).
111. *Hollaender, A.* (editor) Chemical Mutagens, p. 100: New York, Plenum, 1971, Volume 1.
112. *Neunhoeffer, O.:* Z. Naturforsch. B, *25*, 299 (1970).

Received December 27, 1971

Structure and Bonding: Index Volume 1-11